ASCE Manuals and Reports on Engineering Practice No. 134

Water Pipeline Condition Assessment

Prepared by the
Task Committee on Water Pipeline Condition Assessment

Edited by
George F. Ruchti Jr.

Sponsored by the
Pipeline Division of the
Utility Engineering and Surveying Institute of the
American Society of Civil Engineers

ASCE AMERICAN SOCIETY OF CIVIL ENGINEERS

Published by the American Society of Civil Engineers
Reston, Virginia

Library of Congress Cataloging-in-Publication Data

Names: Ruchti, George F., editor. | Utility Engineering and Surveying Institute (American Society of Civil Engineers). Task Committee on Water Pipeline Condition Assessment, author. | Utility Engineering and Surveying Institute (American Society of Civil Engineers). Pipeline Division, sponsoring body.
Title: Water pipeline condition assessment / prepared by the Task Committee on Water Pipeline Condition Assessment ; edited by George F. Ruchti Jr. ; sponsored by the Pipeline Division of the Utility Engineering and Surveying Institute of the American Society of Civil Engineers.
Description: Reston, Virginia : The American Society of Civil Engineers, [2017] |
Series: ASCE manuals and reports on engineering practice ; no. 134 | Includes index.
Identifiers: LCCN 2017015211 | ISBN 9780784414750 (soft cover : alk. paper) |
ISBN 9780784480588 (PDF) | ISBN 9780784480656 (ePUB)
Subjects: LCSH: Water-pipes–Testing. | Water-pipes–Evaluation.
Classification: LCC TD491 .W378 2017 | DDC 621.8/672–dc23 LC record available at https://lccn.loc.gov/2017015211

Published by American Society of Civil Engineers
1801 Alexander Bell Drive
Reston, Virginia 20191-4382
www.asce.org/bookstore | ascelibrary.org

Errata: Errata, if any, can be found at https://doi.org/10.1061/9780784414750.

ISBN 978-0-7844-1475-0 (print)
ISBN 978-0-7844-8058-8 (PDF)
ISBN 978-0-7844-8065-6 (ePUB)
Manufactured in the United States of America.

24 23 22 21 20 19 18 17 1 2 3 4 5

Cover photo: Courtesy of Pure Technologies U.S.

MANUALS AND REPORTS ON ENGINEERING PRACTICE

(As developed by the ASCE Technical Procedures Committee, July 1930, and revised March 1935, February 1962, and April 1982)

A manual or report in this series consists of an orderly presentation of facts on a particular subject, supplemented by an analysis of limitations and applications of these facts. It contains information useful to the average engineer in his or her everyday work, rather than findings that may be useful only occasionally or rarely. It is not in any sense a "standard," however; nor is it so elementary or so conclusive as to provide a "rule of thumb" for nonengineers.

Furthermore, material in this series, in distinction from a paper (which expresses only one person's observations or opinions), is the work of a committee or group selected to assemble and express information on a specific topic. As often as practicable the committee is under the direction of one or more of the Technical Divisions and Councils, and the product evolved has been subjected to review by the Executive Committee of the Division or Council. As a step in the process of this review, proposed manuscripts are often brought before the members of the Technical Divisions and Councils for comment, which may serve as the basis for improvement. When published, each work shows the names of the committees by which it was compiled and indicates clearly the several processes through which it has passed in review, so that its merit may be definitely understood.

In February 1962 (and revised in April 1982), the Board of Direction voted to establish a series titled "Manuals and Reports on Engineering Practice," to include the Manuals published and authorized to date, future Manuals of Professional Practice, and Reports on Engineering Practice. All such Manual or Report material of the Society would have been refereed in a manner approved by the Board Committee on Publications and would be bound, with applicable discussion, in books similar to past Manuals. Numbering would be consecutive and would be a continuation of present Manual numbers. In some cases of joint committee reports, bypassing of Journal publications may be authorized.

A list of available Manuals of Practice can be found at http://www.asce.org/bookstore.

CONTENTS

PREFACE

This Manual of Practice provides an investigator with an appropriate path to adequately determine the structural status of an individual pipeline or pipeline system. The manual provides recommendations and information to assist in a detailed approach to prioritize a program and develop an accurate determination of the condition of the individual pipelines. With an understanding of the structure of such a program, understanding the nature of the materials to be evaluated, the tools available for the evaluation, and what information the findings should provide, the Water Pipeline Condition Assessment report can and should provide critical input to an asset management program.

The "Introduction" walks the reader through the steps to achieve this goal and provides the framework for this manual. The next chapters provide information on the type of reporting that utilities should be implementing and the planning that then goes into the structure of the condition assessment. Six chapters, written by experts in each material, discuss pipeline materials including cast iron, ductile iron, concrete (all variations), steel, fiberglass, PVC, and asbestos cement pipe. Although repair, rehabilitation, or replacement of the individual materials is not a function of the condition assessment, they are an important part of an asset management program; therefore, information on each product has been provided for assistance. "Pipe Inspection Tools" deals with the various tools available for the assessment, for which materials they can be used, and what they can and cannot accomplish. The Condition Assesment Report chapter provides an understanding of the results that should be expected, and the "Case Studies" provide guidance from actual projects covering the majority of the pipe products that have been available.

ACKNOWLEDGMENTS

Task Committee on Water Pipeline Condition Assessment

Chair: George F. Ruchti Jr., M. ASCE, Lockwood, Andrews & Newnam
Vice Chair: John J. Galleher, P.E., Pure Technologies U.S.

Chapter Chairs

Chapter 1: Dan Carroll, P.E., Aurora Water
Chapter 2: Randy Randolph, P.E., Central Arizona Project
Chapter 3: John J. Galleher, P.E., Pure Technologies U.S.
Chapter 4: L. Gregg Horn, P.E., M. ASCE, Ductile Iron Pipe Research Association
Chapter 5: Timothy Kennedy, P.E., Consultant
Chapter 6: Brent Keil, P.E., Northwest Pipe Company
Chapter 7: C. Douglas Jenkins, P.E., CH2M
Chapter 8: Steve Cooper, P.E., Uni-Bell PVC Pipe Association
Chapter 9: Britt Klein, Tucson Water
Chapter 10: Mark Holley, Pure Technologies U.S.
Chapter 11: Jack Adam, P.E., KEH and Associates
Chapter 12: Ken Wilson, P.E., Atkins Engineers

Members

Sam Arnaout, P.E., Forterra
Graham Bell, Ph.D., P.E., HDR
Randy Conner, P.E., American Cast Iron Pipe Company
Nathan Faber, P.E., San Diego County Water Authority
Michael Fleury, P.E., BCEE, Carollo Engineers
Ahmad Habibian, Ph.D., P.E., CDM Smith
Bill Heubach, P.E., Seattle Public Utilities
Tom Iseley, Ph.D., P.E., Dist. M. ASCE, PWAM, Louisiana Tech

Dan Daehyun Koo, Ph.D., P.E., Indiana University-Purdue University-Indianapolis
Jeffrey LeBlanc, P.E., Thompson Pipe Group-Flowtite
Dale Lindemuth, P.E., Corrpro Companies
David Marshall, P.E., Tarrant Regional Water Authority
Richard I. Mueller, P.E., F.ASCE, American Concrete Pressure Pipe Association
Rasco Ojdrovic, Ph.D., P.E., Simpson Gumpertz & Heger
Rafael Ortega, P.E., Aurora Technical Services
Edward A. Padewski III, P.E., Pure Technologies U.S.
John Plattsmier, P.E., HDR
Camille George Rubeiz, P.E., F.ASCE, Plastics Pipe Institute
Jose L. Villalobos, P.E., V&A Consulting Engineers
Andrew Williams, P.E., Louisville Water Company

Blue Ribbon Panel

John Bambei
John Economides
Michael Stift

CHAPTER 1
INTRODUCTION

Various asset management programs can be employed to maximize the useful life of a water utility's infrastructure. An asset management program establishes a baseline to understand and anticipate present and future use of resources. In addition, the Government Accounting Standards Board (GASB) monitors and establishes reporting methods for state and local governments. Statement No. 34, also known as GASB34 provides two methods for reporting a water utility's assets:

1. Reporting of infrastructure assets is calculated by the historic cost less depreciation model.
2. Reporting is done by the infrastructure asset management system, also known as the Modified Approach, and is based upon the utility meeting the following requirements:
   a. Have an up-to-date inventory of eligible infrastructure assets.
   b. Perform condition assessments of the eligible infrastructure assets every three years and summarize the results using a measurement scale.
   c. Estimate each year the annual amount to maintain and preserve the eligible infrastructure assets at the condition level established and disclosed by the water utility.

Although more costly to provide, the Modified Approach offers tax incentives to a private water utility. This is a further incentive for a private utility to employ an asset management program.

This document will focus on the management of the transmission pipeline portion of a water utility's infrastructure, not the distribution system piping, as it is considered an ongoing maintenance issue. The condition assessment of a pipeline is an asset management tool that is being employed by more utilities rather than wholesale replacement of troublesome pipelines. The guidelines presented in this document address the condition assessment of the pipeline itself. It does not address the assessment of pipeline appurtenances, such as control valves, blowoff valves, hydrants, air or vacuum valves, etc., unless they were determined to be at fault during the condition assessment and then such appurtenances should be duly noted. It is anticipated that these appurtenances will be exercised or inspected on a regular schedule and repaired or replaced as necessary.

1.1 CONDITION ASSESSMENT

The condition assessment of a pipeline is the evaluation of its current state and its associated components with respect to its ability to meet the defined needs of its owner. It may include an evaluation of its state of deterioration, its ability to withstand anticipated loads, its state of water tightness, and its level of hydraulic capacity. Condition assessment must also be time dependent; that is, it must address not only the present state of a pipeline but also make some determination of its remaining service life. There are few absolutes associated with a condition assessment program or the results of such a program, but there is one absolute in not pursuing a program, and that is the eventual loss of functionality of the system and increased risk to the public.

The service life of an asset is the time period the asset meets the defined needs of the owner. The current defined needs may be different from the original. Remaining life estimates usually result in very broad forecasts, such as

- The asset replacement or decommissioning is required as soon as possible;
- Over the next 20 years with normal maintenance, the asset is predicted to meet the owner's needs; or
- Major maintenance is required within the next 10 years for the asset to meet the owner's needs.

Deterioration may occur on the exterior surface, the interior surface, or both. It may take the form of oxidation of a metallic surface or a chemical attack. It may be that the product had a structural defect or sustained damage. The pipeline may currently be leaking or it may demonstrate a limiting term of continued water tightness and may even be in a state of structural distress or imminent failure. While oxidation or rusting of the interior surface of a metallic pipeline may or may not pose leakage issues, significant tuberculation hampers flow properties and reduces the pipeline's hydraulic capacity. Another concern, possibly to a lesser degree, is when the water quality leaches cement or is detrimental to the aggregate in a cement mortar or concrete lining.

Condition assessment is an important tool for any asset management program. To effectively assess the condition of a pipeline or a system of pipelines, an inventory of the materials is critical to the success of the program. This inventory should include, but not be limited to pipe material, manufacturer, age, lining, coating, type of joint, maintenance history, failure history, construction history, design reports, inspection reports, and future expectations for the pipeline (i.e., does it meet the expected demand of the utility?). For the purposes of this document, a failure is considered to be a joint leak, a barrel leak, a rupture, or a collapse of a pipeline. To a lesser extent, as such information could be determined by the condition assessment team, it would be helpful to have data on soil corrosive properties and soil contamination and any known issues with stray currents from nearby or crossing pipelines, light rail tracks, high voltage power lines, or any other source of stray currents. While a failure is certainly not beneficial, if a pipeline should fail, the utility should take full advantage of any data that could be applied to other areas of the system to mitigate any failures in the future.

A buried water pipeline poses a considerable level of difficulty to determine the exact condition and how much service life remains. Uncovering portions of the pipeline only provides a snapshot of the line, and it may or may not be an accurate indicator of the entire pipeline's condition.

Companies have developed electromagnetic devices to investigate the structural integrity of certain prestressed concrete water pipeline materials. These devices, in conjunction with the more permanent incorporation of acoustic monitoring technology, can provide present and time dependent assessment of the pipeline's condition.

Tools have been developed by the oil and gas industry to evaluate the condition of buried metallic pipelines. However, these tools are not effective for assessing the condition of plastic or other nonmetallic pipeline materials employed by the water industry.

For many metallic pipelines, particularly those with bonded coatings, an over the line pipe-to-soil survey can be very effective in determining corrosive conditions that may then be quantified with some evaluation of time dependency. This procedure eliminates, or at least reduces, the impact of excavations. However, some types of installation pose greater levels of difficulty in detecting those areas of concern, such as under pavements and in congested areas.

Pipeline leakage is an indication of its condition and can now be evaluated by several means including hydraulic modeling and measurements and acoustic monitoring. With the advent of several acoustical mechanisms that are intrusive to the system, it is possible to detect leakage and approximate leak locations and leakage rates.

Most of the foregoing discussion regarding deterioration concerns metallic pipeline material. Fiberglass, asbestos, and plastic materials have a different set of concerns. Permeation of the wall, chemical attack, physical damage (cracks, scratches, and gouges), leaks, joint imperfections, age embrittlement, and nonelastic pipe deformation can pose concern for the long-term service of such pipelines.

Naturally, the more thorough and exhaustive the condition assessment, the more expensive it is. Pipeline owners do not have unlimited budgets. Compromise must therefore be made in developing the assessment program. The criticality of the pipeline, its failure history, its redundancy, the degree and depth of the investigation, etc., must be evaluated for maximizing cost effectiveness.

These guidelines are a result of the collaboration of a multitude of industry experts with a combined experience of more than 874 years. The assembled knowledge was acquired through material research and the implementation of previous condition assessment projects by various utilities. This information should be used to assist utilities with making informed decisions in initiating a comprehensive pipeline condition assessment program. These guidelines presented in 12 chapters, examine seven important topics:

- Introduction,
- Data collection and management,
- Planning,
- Pipe materials,
- Pipe inspection tools,
- Condition assessment report, and
- Case studies.

Important data collection and management records, which aid in completing water pipe condition assessments, are the routine reports completed after both an average water pipe leak repair and a critical or catastrophic failure. These reports provide a basis for evaluating the condition of water pipes. Suggested report data are provided for basic leak repairs, for advanced reporting, and for data provided by a utility administrator. Two sample forms are provided that highlight the suggested reporting data and format.

The "Planning" chapter presents ideas for managers to consider before initiating a pipe condition assessment. Also presented are two styles of condition assessment: function based and risk based. The chapter concludes by providing an example of general criteria for water line ranking.

Chapters 4 through 9 discuss six water pipe materials:

- Cast iron and ductile iron,
- Concrete,

- Steel,
- Reinforced fiberglass (also referred to as fiber-reinforced plastic),
- PVC, and
- Asbestos cement

Each of these chapters begins with a history of the pipe material before discussing the characteristics and circumstances that affect the condition of each particular pipe material. Techniques for completing condition assessments are presented along with the overall determination of the pipe condition. Each of these chapters concludes by discussing rehabilitation practices for each pipe material. This is not information to be incorporated in the final condition assessment report. Rather, rehabilitation and replacement have been included for information purposes only and are intended to provide an understanding of available options, particularly should a situation of eminent product failure be discovered during the assessment.

There are other pipe materials that are or have been used in water pipe service but providing a definitive condition assessment of their structural integrity or external surface condition is difficult, either because of their age and availability, primary installation applications, the extreme difficulty of assessing their condition, or because of their mode of failure.

The "Pipe Inspection Tools" chapter provides a list of 50 tests that can be used to evaluate the condition of water pipe, depending on the pipe material. These tests are listed in a joint Water Environment Research Foundation and Water Research Foundation report for condition assessments of water and wastewater utility assets. The list details the pipe materials for which each test is applicable.

The "Condition Assessment Report" chapter provides a discussion about what to do with the pipe testing results. Classification of these results allows the utility to evaluate the complete system of water pipes and determine where the pipes are in their life cycle and level of service. An example system is provided to illustrate a numerical value assignment that summarizes condition assessment results.

The final chapter presents case studies, summarizing condition assessment projects completed by the following utilities:

- Weber Basin Water Conservancy District in Utah,
- Miami-Dade Water and Sewer Department in Florida,
- Tarrant Regional Water District in Texas,
- San Diego County Water Authority in California,
- Pinellas County Utilities in Florida,
- Dallas Water Utilities in Texas, and
- City of Houston Department of Public Works and Engineering in Texas.

These case studies were completed on five different pipe materials. Details presented include planning, inspection strategies, tools and inspection methods used, condition assessment results, and recommendations.

One important note: Particular attention should be paid to avoid potential surges when refilling or repressurizing a line that has been shut down for any reason. Also, the owner should keep in mind the importance of future proper maintenance of their pipelines and appurtenances.

1.2 PERTINENT WEBSITES

American Petroleum Institute (API): http://www.api.org/
American Public Works Association (APWA): http://www.apwa.net/
American Society of Civil Engineers (ASCE): http://www.asce.org/
American Society of Mechanical Engineers (ASME): http://www.asme.org/
American Society for Testing and Materials (ASTM): http://www.astm.org/
American Water Works Association (AWWA): http://www.awwa.org/
Ductile Iron Pipe Research Association (DIPRA): http://www.dipra.org/
EPA Aging Water Infrastructure Research (AWI): http://www.epa.gov/awi/index.html
EPA National Service Center for Environmental Publications (NSCEP): http://www.epa.gov/nscep/index.html
Fiberglass Tank & Pipe Institute: www.fiberglasstankandpipe.com
Government Accounting Standards Board (GASB): http://www.gasb.org/
National Research Council (NRC): http://www.nationalacademies.org/nrc/index.html
National Research Council Canada (NRCC): http://www.nrc-cnrc.gc.ca/eng/index.html
National Sanitation Foundation (NSF): http://nsf.org/
Occupational Safety and Health Administration (OSHA): http://www.osha.gov/
Plastic Pipe Institute (PPI) / HDPE: http://plasticpipe.org/
PVC Pipe Association: http://www.uni-bell.org/
United States Army Corps of Engineers, Engineer Research and Development Center: http://www.erdc.usace.army.mil/
Water Environment Federation (WEF): http://wef.org/
Water Environment Research Foundation (WERF): http://www.werf.org/
Water Infrastructure Database: http://waterid.org/
Water Research Foundation (WRF): http://www.waterrf.org/Pages/WaterRFHome.aspx

West Virginia University, National Environmental Services Center (NESC): http://www.nesc.wvu.edu/
WRc: http://www.wrcplc.co.uk/home.aspx

1.3 RELATED DOCUMENTS

Aging Water Infrastructure Research Program (2004)
EPA/600/R-09/049: http://www.epa.gov/nrmrl/pubs/600r09055/600r09055.pdf
Water Research Foundation (2002)
Water Research Foundation (2008)

1.4 ACRONYMS

AC	asbestos cement
AE	acoustical emissions
AFO	acoustic fiber optic
AI	aggressive index
API	American Petroleum Institute
APWA	American Public Works Association
ARV	air release valve
ASME	American Society of Mechanical Engineers
AWI	aging water infrastructure
AWWA	American Water Works Association
BTEX	benzene, toluene, ethyl benzene, xylene
BWZ	broken wire zone
CA	condition assessment
CC	concrete cylinder
CCTV	closed-circuit television
CFRP	carbon fiber reinforced polymer
CI	cast iron
CIOD	cast iron outside diameter
CIP	capital improvement program
CMMS	computerized maintenance management system
CP	cathodic protection
CSIRO	Commonwealth Scientific and Industrial Research Organization (Australia)
DI	ductile iron
DIPRA	Ductile Iron Pipe Research Association
DSC	differential scanning calorimetry
DWU	Dallas Water Utilities
EC	embedded cylinder

EDS	energy dispersive spectroscopy
EM	electromagnetic
ESCR	environmental stress-cracking resistance
FRP	fiberglass-reinforced pipe
GASB	Government Accounting Standards Board
GIS	geographic information system
GRP	glass-reinforced plastic
HDPE	high-density polyethylene
LI	Langelier index
MCL	maximum contamination level
MFL	magnetic flux leakage
MSDS	Material Safety Data Sheet
NACE	National Association of Corrosion Engineers
NDE	nondestructive examination
NDT	nondestructive testing
NESC	National Environmental Services Center
NPS	National Pipe Standard
NRC	National Research Council
NRCC	National Research Council Canada
NSCEP	National Service Center for Environmental Publications
NSF	National Sanitation Foundation
OSHA	Occupational Safety and Health Administration
PCCP	prestressed concrete cylinder pipe
PE	polyethylene
PENT	polyethylene notch tensile test (ASTM 1473)
PPI	Plastic Pipe Institute
PPIC	Pressure Pipe Inspection Company
PPR	pipe penetrating radar
RC	reinforced concrete
RCP	rapid crack propagation
RCT	risk curves technology
RFEC	remote field eddy current
RPMP	reinforced polymer mortar pipe
RTRP	reinforced thermosetting resin pipe
SCG	slow crack growth
SDCWA	San Diego County Water Authority
SEM	scanning electron microscope
SFPUC	San Francisco Public Utilities Commission
SGH	Simpson, Gumpertz & Heger
SSDA	Second San Diego Aqueduct
TC	transformer coupling
TRWD	Tarrant Regional Water District
UKWIR	United Kingdom Water Industry Research
USACE	United States Army Corps of Engineers

UT	ultrasonic testing
WBWCD	Weber Basin Water Conservancy District
WEF	Water Environment Federation
WERF	Water Environment Research Foundation
WIRP	water infrastructure replacement prioritization
WRc	Water Research Centre
WRF	Water Research Foundation

CHAPTER 2

DATA COLLECTION AND MANAGEMENT

As will be discussed in the "Planning" chapter, pipe condition data collection and management is an important tool for water utilities, providing they offer accurate and valuable information. The information must be comprehensive enough to allow a utility to make informed decisions regarding the quality of pipeline assets during the condition assessment process. Each utility can customize the type of information needed to meet the needs of management, staff, and others. The data are of little value if not easily retrievable. A pipeline system may be fairly simple, which allows documents to be tracked using spreadsheets, or quite complex, in which case the use of a document management system, geographic information system (GIS), or other database is more beneficial.

2.1 DATA MANAGEMENT

Management of data for a pipeline as well as an agency's entire pipeline program will need to be collected and managed over the life of the pipeline. The data should be organized by pipe section. Each stick of pipe should have baseline data (e.g., mark number, material, manufacture date, class of pipe, etc.) as well as every bit of information gathered about it over its life (e.g., cathodic protection test data, leaks, repairs, inspection data, maintenance information, corrosion assessments, captioned photos, and details of other construction that has occurred around it such as a utility crossing). As mentioned in the introductory section of this chapter, a document management system, GIS, or other database may be necessary to store data in a manner that is retrievable and easily utilized. Many agencies are now embracing technology and use tablets, handheld devices, smart phones, global positioning system (GPS) devices, and other readily available technologies as part of their overall pipeline management program.

2.2 HISTORICAL DOCUMENTS

To adequately manage pipeline systems, the retrieval and assembly of available historical records are required. These records provide an inventory of pipeline assets along with baseline information necessary to begin a pipeline condition assessment program. If records for a utility's entire water distribution system are collected, reviewed, and stored as part of the condition assessment process, it will be much easier to manage the system going forward. These documents provide much needed data that can be used when planning and scheduling future work, identifying trends, calculating pipe deterioration rates, determining maintenance/repair intervals, and estimating remaining useful life. In addition, when gaps in data are identified, it may be possible to get the information from staff, manufacturers, contractors, or consultants. If required data are missing, plans to obtain or supplement these gaps can be developed.

Table 2-1 provides a summary of documents that are sometimes available in the utility's archives to initiate this process.

2.3 CONDITION ASSESSMENT REPORTS

Pipeline condition assessment is an important part of a water utility's asset management program. Each time a pipeline is investigated, whether for a planned inspection, condition assessment testing, or a failure (such as a leak or ruptured pipe section), a report should be prepared. Report complexity can vary from a routine checklist to a full root cause failure analysis. The type of report required is dependent upon the needs of the utility, criticality of the asset, the funds available, time availability, and the report audience.

Condition assessment results can be incorporated into the utility's asset management program or GIS. These results can be used to

- Assess the pipeline infrastructure's ability to meet current and future water delivery requirements;
- Identify and evaluate optimal strategies to maintain infrastructure, to repair defects, and/or to rehabilitate/replace pipelines; and
- Estimate rate of deterioration/degradation by comparing multiple reports on the same pipe system over a period of time.

Table 2-1. Historical Documents for Pipeline Condition Assessment

Documents	Description	Purpose
Plans and specifications	Provide contract documents, drawings, and specifications	May define pipe material options; identify right-of-way available, general layout, trenching, bedding, and handling requirements
Contract modifications	Provide records of changes that were required during construction	Provide indication of design deficiencies and field conditions
Project final report and record drawings	Define as-built conditions at completion of construction	Often include details such as location of compaction tests, soil conditions/classification, trench stability, pipe foundation, backfill, location of utilities, cathodic protection details, and other relevant information
Pipelaying diagrams	Identify pipe numbering, pipe material, diameter, lay sequence, class (pressure and cover), fittings, valve locations, thrust restraint, and detailed layout	Define pipe type, plan, and profile
Shop drawings	Provide details for manufacturing each pipe section	Provide supplier and manufacturing details, joint types, bevels, linings, coatings, and special fittings
Utilities	Provide location of utilities and utility crossings	Prevent damage to all utilities
Right-of-way	Define right-of-way ownership (fee, easement, or license)	Define work limits, contact information for adjacent land owners, permitting, and legal land use rights

(Continued)

Table 2-1. Historical Documents for Pipeline Condition Assessment *(Continued)*

Documents	Description	Purpose
Geotechnical report	Define soil types, depth to rock, depth to water, changes in water depth, soil gradations, soil resistivities, soil corrosivity, and trench slope requirements	May be relevant to the original design, future repair work, and preventive maintenance requirements
Contractor submittals	Provide approved submittals required by the contract	Define materials approved for the project
Inspection reports	Provide inspection reports generated during manufacturing, installation, and throughout the life of the pipeline	Provide dates of manufacture and installation; identify repairs made at the time of construction, field conditions, field methods, test results, fitting location, weather, problems encountered, significant discussions, and personnel on the jobsite each day
Photos	Provide photographs taken and captioned during manufacture and construction	Provide job history and construction methods
Test reports	Provide quality control documentation for items such as welds, materials testing, bedding, backfill, compaction, linings, coatings, cathodic protection, pressure, and leakage	May be used to verify compliance with the design and to compare with current code requirements
Maintenance and repair history	Document repairs completed during manufacture, installation, and throughout the life of the pipeline	May be useful for planning and scheduling maintenance, projecting future maintenance costs, and estimating remaining service life

Pipeline draining reports	Provide standard operating procedures for draining the pipeline	May be helpful for safely draining the system, in particular as staff turnover occurs; defines methods for dewatering and refilling the pipeline including timeframes required; valve, air valve, gate, and pumping sequences; pressures, customer notifications, and clearances required
Pipeline refilling procedures	Provide standard operating procedures for refilling the line	Define methods for safely refilling the system
Safety	Provide job safety analyses, lockout/tagout procedures, ventilation, and tapping	Provide steps necessary for working safely inside a confined space; can be refined for future work but provides a good starting point
Design summary	Define design logic, standards, codes, transient analysis, and other details	Can be used to gain understanding of system and code comparison

Leaks are often discovered by visually seeing water seeping or flowing where it is not a normal occurring condition. Several internal and external leak detection technologies are also available. A sample of a routine leak detection report is shown in Fig. 2-1.

A major leak or the complete failure of a pipeline requires more detail than the report for a routine leak and may also need to address political, legal, and other needs of the agency.

The following shows a sample of an outline for a pipeline failure report:

1. Executive summary
   a. Incident summary
   b. Root cause investigation
   c. Causes and other issues
      (1) Apparent cause
      (2) Root causes
      (3) Contributing factors
   d. Damage assessment
   e. Immediate corrective action
   f. Corrective action to prevent/mitigate recurrence
2. Introduction
   a. Background information
   b. Sequence of events
3. Root cause failure analysis methodology
   a. Design review
      (1) Specifications
      (2) Modifications
   b. Application review
      (1) Procurement
      (2) Installation
      (3) Operation of system
      (4) Vendor's specifications
      (5) Inspection
      (6) Pipe material properties
      (7) Reliability history
      (8) Maintenance history
      (9) Modifications
   c. Identification of root causes
   d. Identification of corrective actions
   e. Findings
4. Damage assessment
   a. Summary of findings
   b. Pipe materials assessment
5. Recommendations
   a. Immediate corrective action
   b. Long-term corrective action

AGENCY

LEAK INSPECTION/REPAIR REPORT

LOCATION ____________

WORK ORDER NO. ____
BLUE STAKE NO. ____
DATE ____

PIPE MATERIAL	JOINT	EXTERNAL PROTECTION	SURFACE CONDITIONS	SOURCE OF REPORT
☐ (CI) Cast Iron ☐ (DI) Ductile iron ☐ (St) Steel ☐ (Cn) Concrete ☐ (FRP) Fiberglass ☐ (HDPE) High Density Polyethylene ☐ (PVC) Polyvinylchloride ☐ (AC) Asbestos Cement	☐ Lead ☐ Rubber push-on ☐ Mechanical ☐ Flanged ☐ Restrained Type ____ ☐ Not seen ☐ Other ____	☐ Bitumen ☐ Plastic wrap ☐ Cathodic protect ☐ None visible ☐ Other ____ INTERNAL PROTECTION ☐ Mortar ☐ Coating ☐ Other ____ ☐ None visible ____	☐ Heavy traffic ☐ Light traffic ☐ Outside of street GROUND TYPE ☐ Soil ☐ Clay ☐ Fill ☐ Rock Bedding? ☐ yes ☐ No	☐ Inspection ☐ Compliant ☐ Other INDICATION OF LEAK ☐ Sonic ☐ Surface Water ☐ Low Pressure ☐ Other

FOREMAN ____________

1. WHAT PART OF MAIN WAS DAMAGED?
a ☐ Pipe barrel d ☐ Flange nuts, bolts, tie rods
b ☐ Joint e ☐ Other (Explain on back)
c ☐ Valve

2. WHAT CAUSED THE DAMAGE IN YOUR OPINION?
a ☐ Ground movement – only if evident
b ☐ Contractor's equipment
c ☐ Coating failure
d ☐ Corrosion/electrolyses
e ☐ Other ____
f ☐ See comments
g ☐ Unknown

3. WHAT TYPE OF CORROSION DAMAGE?
a ☐ No corrosion damage d ☐ Graphitized cast iron (Pipe looks OK, but has little strength because the Iron has dissolved, leaving mostly carbon)
b ☐ Pitting
c ☐ Gen corrosion

4. WHAT REPAIRS WERE MADE?
a ☐ Leak clamp f ☐ Other ____
b ☐ Welded
c ☐ Repacked valve
d ☐ Recaulked joint
e ☐ Replaced section

5. SHOULD THE PIPE BE REPLACED?
a ☐ Yes b ☐ No c ☐ Not sure

6. HOW BIG WAS THE LEAK?
a ☐ Break (entire circumference) c ☐ Large hole
b ☐ Small hole (under 1") d ☐ Split

7. WHAT IS NATIVE SOIL LIKE?
a ☐ Clay or adobe c ☐ Sandy
b ☐ Loam d ☒ Gravel/rock

8. INSTALLED ANODE AND POLYWRAP?
(Omit the anode & Polywrap only on non-metallic pipe)
a ☐ Yes
b ☐ No

9. DEPTH OF COVER IN FT. (Check Closest Number)
10+ ☐ 10 ☐ 9 ☐ 8 ☐ 7 ☐ 6 ☐ 5 ☐ 4 ☐ 3 ☐ 2 ☐ 1 ☐

10. WHERE WAS THE LEAK? (Sketch)

COMMENTS	
ATTACHMENTS	Enter attachment here
PHOTOS	Enter photos here

Fig. 2-1. Sample of a routine leak detection report
Note: HDPE is shown on this sample report, but is not covered in this book.

6. Appendices
   a. Photographs
   b. Supporting documents

In summary, standard pipeline condition inspection and assessment reports should be simple and streamlined to be effective and efficient for use by utilities. Major condition assessment and failure reports generally require much more time, effort, and expense to prepare. In many cases, multiple reports are required to meet the needs of the utility. Reports prepared by the owner, testing company, and consultants are often required to complete the condition assessment process.

Assembling historical data in an organized, retrievable manner prior to a failure event will facilitate a better condition assessment program in the future. The authors hope this manual may provide more insights to many and help the industry create better consensus for pipe inspection and assessment.

CHAPTER 3
PLANNING

When implementing a condition assessment project, it is important to note that no single technology or technique provides a comprehensive assessment of a pipeline. By understanding that many factors impact the performance of pipelines, a realistic assessment of the pipeline can be made. Too often, assessment strategies are implemented without consideration of the particular threats and informational needs required to make a defensible decision for effective pipeline management. By using this *broad brush approach* to condition assessment, valuable resources are often depleted without providing the necessary information for managing the pipeline.

To effectively manage pipelines, an understanding of the pipeline's likelihood of failure (i.e., the pipe's condition) and consequence of failure must be understood to make knowledgeable decisions in the maintenance and extension of the pipeline's remaining useful life (Fig. 3-1). Therefore, the goal of any successful management approach is to provide a cost-effective, reliable, and defensible inspection and assessment strategy that determines the existing condition of the pipeline, estimates the remaining useful life of the asset, and provides long-term capital planning associated with any corrective actions (rehabilitation and/or replacement). To maximize the value of a condition assessment project, two things must occur. First, a cost-effective and reliable assessment approach must be developed that focuses resources on collecting the appropriate condition information for a pipeline. Second, the outcome of the condition assessment must reduce the pipeline's risk of failure.

To develop this cost-effective and reliable assessment approach, a plan must be developed to ensure that the correct information is collected and used to make sure that the pipeline's risk is reduced (primarily through

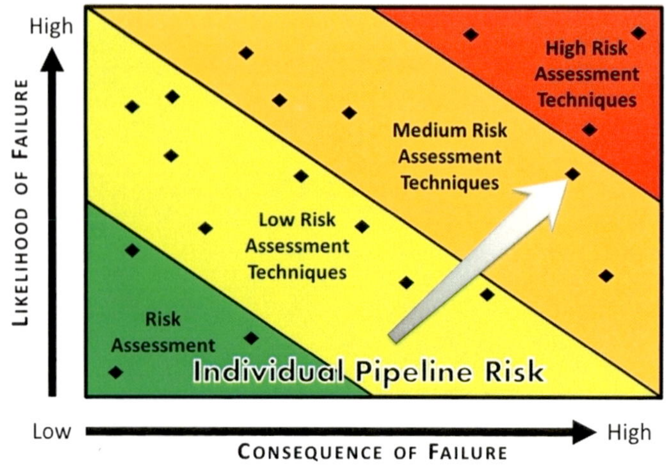

Fig. 3-1. Risk-based condition assessment
Source: Courtesy of Pure Technologies U.S.

reducing the likelihood of failure). Planning for a comprehensive water pipe condition assessment involves three major tasks:

1. Documentation of pipeline assets and data on those assets (What do you have?);
2. Prioritization and scheduling of condition assessments (Which pipes are most critical and are highest priority?); and
3. Planning condition assessment logistics (How are you going to get it done? What tools do you need to make the appropriate decisions?).

3.1 DOCUMENTATION OF PIPELINE ASSETS

Before gathering pipeline information, it helps to understand the factors that may be used for condition assessment prioritization. In many cases, prioritization is based on the following:

1. Location,
2. Age,
3. Material and its properties,
4. Failure history (forensic reports),
5. Surrounding environment,
6. Previous assessment information,
7. Operations,
8. Design,
9. Manufacturing and construction,

10. Maintenance,
11. Failure consequences (e.g., effect on system operation, meeting service levels, and potential third-party damage), and
12. Right-of-way management.

There are many sources of information (data) that can be assembled for each major prioritization category, including

- Pipeline drawings (design or as-built drawings)—*Location;*
- Maps and schematics—*Location;*
- Databases including computer-aided design and geographical information system databases—*Location;*
- Failure records—*Failure history;*
- Soil and geology data such as corrosivity, chlorides, sulfates, and pH—*Surrounding environment;*
- Assessment and inspection records—*Previous assessment information;*
- Operation records such as pressures, transient pressure changes, flow rates, changing operations, or changing water quality—*Operations;*
- System operation information such as normal operation, backup operations, and redundancy—*Operations;*
- Design and construction standards—*Design;*
- Construction specifications—*Manufacturing and construction;*
- Manufacturing, construction, and inspection records—*Manufacturing and construction;*
- Maintenance records such as repairs, inspections, and cathodic protection—*Maintenance;* and
- Easement information such as whether it is owned in fee, adjacent utility easements, limits and utility crossings, permits required—*Right-of-way management.*

Some information may be readily available, and some may be difficult to obtain or not available at all. If limited data are available, initial prioritization and data gathering efforts may only focus on one or two factors. In some systems, there may be only one or two conditions that contribute to their failures. It may be the age of the pipe or the corrosiveness of the soil. Also, utilities should consider the information that is required for the condition assessment effort including basic information such as the location of the pipeline, the pipeline material, and specifications. Data gathering can be extremely time consuming, so it is important to focus on data that are available or easily gathered and consider additional data collection for future efforts.

3.2 PRIORITIZATION AND SCHEDULING

The vast majority of utilities do not have the resources to universally assess the condition of their water pipeline assets. Consequently, they must develop a plan to sequence those assets through a prioritization schedule.

The plan must be individualistic, based on the nature of the utility and the data available. Is it large or small, urban or rural? Is it an old system or a relatively new one? Is it a retailer or primarily a wholesaler? Is its failure history reflected in its records? Based on these and other answers, this plan could be very simplistic to highly complex. In the case of a small utility, the plan may be intuitive, dependent on the experience or concerns of its knowledgeable management. In the case of a large, complicated system, the plan may entail a complex matrix of conditions to formulate the assessment prioritization.

A condition assessment plan may be function based or risk based. In the case of a function-based plan, the prioritization emphasis is placed on the likelihood of failure. Likelihood of failure factors may include

1. Failure history,
2. Age,
3. Surrounding environment (e.g., soil aggressiveness/corrosivity),
4. Condition,
   - Previous assessment information
   - Remaining service life
   - Decay index
   - Replacement year
5. Material and its properties,
6. Manufacturing and construction,
7. Design (factor of safety),
8. Operations,
9. Maintenance (computerized maintenance management system records check),
10. Location (third-party damage, additional loads), and
11. Aggressiveness of conveyed water.

If the prioritization is function based, the prioritization may be simple. For example, if the criteria are based on age, then the oldest pipe or some combination of age and another factor would be selected to be assessed first. If the criteria are based on soil corrosiveness, then the pipeline located in the most corrosive soil would be assessed first. Table 3-1 is an example of a simple function-based prioritization.

A risk-based assessment plan would also take into account the consequences of failure in addition to the likelihood of failure. Risk is the consequence of failure times the probability of failure:

$$\text{Risk} = \text{Consequence of Failure} \times \text{Probability of Failure}$$

Table 3-1. Example of Function-Based Prioritization

Pipeline	Reach/station	Pipe material	Year installed	Age (years)	Soil corrosivity	Priority based on age
A	1	Cast iron	1955	57	Moderately corrosive	1
A	2	Cast iron	1959	53	Moderately corrosive	2
A	3	Cast iron	1962	50	Moderately corrosive	3
B	10+00 to 41+10	Steel	1971	41	Very corrosive	4
B	41+10 to 100+12	Steel	1975	37	Very corrosive	5
B	100+12 to 110+87	PVC	2003	9	Very corrosive	7
C	Main street	Reinforced concrete	1981	31	Corrosive	6

Source: Courtesy of Pure Technologies U.S.

In a risk-based plan, the consequence of failure plays a major role. Factors may include

1. Pipeline location with respect to structures, parks, schools, hospitals, etc. (land use);
   - Downstream drainage
   - Adjacent utilities' impacts
2. Pipe diameter (capacity/amount of water that will be discharged);
3. Pipeline criticality;
4. System redundancy;
   - Water delivered
   - Automatic control versus manual operation
   - Service interruptions
5. Life and safety;
6. Ability to deliver water; and
7. Bypass or backup system (or connection through nearby pressure zone or another water agency).

For example, a pipeline that provides the sole source to a community would take precedence over a looped pipeline system. A pipeline providing water to a hospital or other emergency operation would take precedence over a community distribution line. Each utility needs to determine not only what criteria should be included but what weight to place on each criterion. Table 3-2 is an example of a simple quantitative risk-based prioritization.

The risk score is the probability of failure score multiplied by the consequence of failure score. The highest priority pipeline for condition assessment would be the one with the highest risk score. The index is a relative number assigned to a condition or activity on the pipeline such as age, material, type of environment, or transported product. The value of the index is not always the same between agencies as its impact may not be the same at the different agencies. Weight is how much that index impacts the safe and reliable operation of the pipeline.

Using the data (Table 3-2), the risk score for Pipe C would be calculated as follows:

$$\text{For the probability: } [(\text{Age})\, 50 \times 50\%] + [(\text{Material})\, 28 \times 50\%] = 39$$

$$\text{For the consequence: } [(\text{Nearby Utility})\, 14 \times 50\%] + [(\text{Capacity})\, 96 \times 50\%] = 55$$

So, the risk is calculated using the following:

$$\text{Risk} = (\text{Probability})(\text{Consequence})$$

$$\text{Pipeline C Risk} = (39)(55)/100 = 21$$

Table 3-2. Example of Risk-Based Prioritization

	Probability of Failure					Consequence of Failure					
	Age		Material			Nearby Utility		Capacity			
Pipe	Index	Weight	Index	Weight	Score	Index	Weight	Index	Weight	Score	Risk Score
A	52	50%	10	50%	31	10	50%	100	50%	55	17
B	54	50%	20	50%	37	12	50%	100	50%	56	21
C	50	50%	28	50%	39	14	50%	96	50%	55	21
D	44	50%	40	50%	42	34	50%	70	50%	52	22
E	38	50%	50	50%	44	36	50%	70	50%	53	23
F	10	50%	82	50%	46	76	50%	70	50%	73	34
G	30	50%	40	50%	35	78	50%	10	50%	44	15

Source: Courtesy of Pure Technologies U.S.

Table 3-3. Qualitative Risk Assessment Matrix

	Consequence				
Failure probability	Insignificant	Minor	Moderate	Major	Severe
Almost certain	Moderate	High	High	Extreme	Extreme
Likely	Moderate	Moderate	High	High	Extreme
Possible	Low	Moderate	Moderate	High	High
Unlikely	Low	Low	Moderate	Moderate	High
Rare	Low	Low	Low	Moderate	Moderate

Source: Courtesy of Pure Technologies U.S.

The risk is divided by 100 to simplify the number. If there are not enough data or resources to develop a quantitative risk score, a qualitative risk table may also be used, such as exemplified in Table 3-3. For instance, if a pipeline that is in good condition is next to a hospital, the calculation may look like

Failure probability = Rare
Consequence of failure = Severe
Risk would equal Extreme

3.3 PLANNING CONDITION ASSESSMENT LOGISTICS

After the priority pipelines have been selected, the condition assessment method will need to be identified. Subsequent chapters will help identify the appropriate method based on several factors. In addition to overall planning, several logistical issues associated with a pipeline condition assessment must be considered. Some issues that should be considered for detailed planning are

- Schedule,
- Detailed maps,
- Engineering drawings and as-built data,
- Pipeline draining and filling,
- Internal and external pipeline access and entry,
- Disinfection procedures,
- Pipeline shutdowns,
- Service interruptions,
- Pipeline isolation from the system,
- Cost estimates,
- Personnel responsibilities,
- Equipment required,
- Traffic,
- Safety,
- Requirements for confined space entry requirements,
- Traffic control,
- Public notification,
- Communication procedures,
- Environmental procedures,
- Previous inspection data availability, and
- Available water storage.

A detailed plan, addressing each of the preceding items, should be completed before the condition assessment and made available to all personnel working on the condition assessment effort.

CHAPTER 4

PIPE MATERIALS: CAST IRON AND DUCTILE IRON

Because of its longevity and performance, much of the piping still in use for water service in North America is cast iron (CI) and ductile iron (DI) pipe. In fact, more than 600 utilities in the United States and Canada have had gray CI pipelines in continuous service for 100 years or more, and at least 23 utilities have had gray CI mains in continuous service for 150 years. In 1955, modern DI pipe was introduced into the marketplace and continues to provide a high degree of dependability because of its high strength, durability, impact strength, flexible joining systems, and ability to control corrosion economically. Designed and manufactured to the industry's most stringent standards, DI pipe is a robust pipe that resists damage during shipping and handling and, once installed, withstands the most demanding operating conditions including water hammer; freeze-thaw cycles; deep trench loadings; areas of high water table and heavy traffic; river crossings; pipe on supports; rocky trenches; and areas of shifting, expansive, and unstable soils.

Although both gray and DI pipe can provide long trouble-free service lives, there are iron pipelines in need of either rehabilitation or replacement. This can be the result of an increased frequency of leaks and failures to either the pipe wall or joints. Older, unlined pipe can be a cause for red water complaints and reduced hydraulic capacity due to internal tuberculation. The performance issues associated with iron pipe vary depending on the type of pipe (pit cast gray iron, centrifugally cast gray iron, or DI), the type and constitution of joints, the type of lining (bitumen, cement mortar, etc.), the specific environment (internal loads, external loads, soil type, bedding conditions, etc.), and in some cases the type of external corrosion controls of the era involved.

The rate of deterioration of an iron pipe water system is not simply a function of the pipe age but rather relates to the specific environment acting upon each segment of the system. As a result, simple methods (material type, pipe age, pipe size, joint type, lining type, corrosion protection system, etc.) of determining infrastructure replacement or rehabilitation schedules cannot be used to yield the most economical result of available resources. Instead, a condition assessment protocol must be developed and put into action. Such an ongoing program will allow actual knowledge of the condition and environment affecting each piece of the water system to be accumulated and analyzed. Only then can informed decisions be made with regard to proposed capital expenditures and rehabilitation methods.

4.1 HISTORY

Human ability to cast pipe probably developed from, or coincidentally with, the manufacture of cannon, which is reported as early as the year 1313. There is an official recording of CI pipe manufactured at Siegerland, Germany, in 1455 for installation at the Dillenburg Castle. In the year 1664, King Louis XIV of France ordered the construction of a CI pipe main extending 15 mi from a pumping station at Marly-on-Seine to Versailles to supply water for the fountains and town. Sections of that CI pipeline are still in service on the palace grounds, more than 350 years after they were installed.

The use of gray CI pipe was introduced into the United States in (or shortly before) 1816, and perhaps as early as 1804. The first gray iron pipe was molded in wooden frames or flasks, packed with sand sufficiently moist to retain its shape, and cast in a horizontal position or *on the side*. As the standard length of pipe increased from approximately 3 ft to 6, 9, and 12 ft, the position of the casting molds changed from horizontal to a slope and, finally, to a vertical position by the year 1850, approximately. The centrifugal casting of gray iron pipe in sand-lined or metal molds began in approximately 1925, revolutionizing the manufacturing process; although some pit casting was not discontinued for many years thereafter.

The advent of DI pipe in 1948 was another extremely significant development in the pressure pipe industry. DI pipe was introduced to the market in 1955 and had a period of overlap until it replaced gray CI pipe altogether in the 1960s and 1970s. DI pipe supplanted gray CI pipe primarily because of its dramatically improved structural properties. Both DI and gray iron contain similar amounts of carbon. In gray CI, most of the carbon is present in the form of interlocking flakes of graphite. In DI, an additional step in the treatment of the molten metal during manufacture causes the graphite to take the form of spheroids or nodules. Due to this difference, gray CI pipe tends to fail in a brittle fashion, whereas DI pipe does not.

The first gray CI water mains were not coated or lined but were installed just as they came from the molds following cleaning. After many years, it became evident that the interior of the pipe might be affected by certain types of water. The use of protective linings was proposed, and most of the gray CI pipe sold for water works service after approximately 1860 was provided with a hot dip bituminous lining and coating.

In 1922, the first cement mortar–lined gray CI pipe was installed in the water distribution system of Charleston, South Carolina. Since that time many improvements have been made in the production of cement mortar–lined iron pipe. Today, the cement mortar lining is standard for DI pipe. Applied using both the centrifugal process and the projection method, cement mortar linings prevent tuberculation by a phenomenon known as passivation, which results from the high pH lining being in contact with the pipe wall, and ultimately by providing a physical barrier to the water. Cement linings are also smooth, which results in typically high flow coefficients (e.g., Hazen-Williams 140) that are maintained during the life of the pipe.

At least as far back as the 1664 Versailles pipeline, CI pipes were made with flanged joints. The Versailles pipeline used a combination of lead and leather gaskets. Over the years, joint technologies have evolved, culminating in modern joints that use elastomer gaskets. Flanged DI pipe is available today for many aboveground plant installations and other specialized applications. The bell and spigot joint was developed in 1785 and was extensively used until the 1950s. It was originally assembled by caulking yarn or braided hemp into the base of the annular bell cavity and then pouring molten lead into the remaining space inside the bell. Several types of lead substitutes, such as leadite, were also developed and used from the 1920s to the 1940s. The rubber-gasketed mechanical joint was developed for gas industry use in the late 1920s and continues to be used in the water industry today, primarily as the preferred joint for fittings. The roll-on joint was developed in 1937 and was used for approximately 20 years before its manufacture was discontinued. Assembly of this joint involved a compressed rubber gasket rolled under a restriction ring, followed by caulked square braided jute. The remainder of the joint was packed with a bituminous compound.

The push-on joint was developed in 1956 and represented another important advancement in the water distribution field. This joint consists of a single rubber gasket placed in a groove that is cast inside of the pipe bell. The joint is made up by applying proper lubrication and pushing the spigot end of the pipe into the bell and past the gasket. The outside diameter of the spigot compresses the gasket, forming a pressure-tight seal. The push-on joint continues to be the primary joint in use today. Other special joints, such as ball and socket, plain end coupled, threaded and coupled, and many proprietary restrained joints, have also been used.

4.2 CONDITION ISSUES

4.2.1 Internal Surfaces

Internal Sedimentation and Scale Formation Under certain conditions sediments can deposit, and in very hard waters internal scales can also form that can clog plumbing and distribution pipes. Both of these phenomena can occur regardless of the pipe materials. Scale is typically composed of calcium carbonate ($CaCO_3$), magnesium hydroxide [$Mg(OH)_2$], calcium sulfate ($CaSO_4$), and/or other compounds. These conditions can be corrected through mechanical cleaning or *pigging* the interior of the pipeline.

Corrosion In certain parts of the United States where aggressive potable waters are encountered, unlined CI water mains may develop growths on the interior surface of the pipe. These growths are called tubercles, and their accumulation has a negative effect upon flow capacity. The inside diameter of the main is reduced, resulting in constricted flow and additional frictional head loss. This tuberculation can also cause water quality and customer service issues due to complaints about red or rusty water. These growths only occur where the water comes in contact with unlined pipe.

Internal corrosion of iron pipe does not typically result in mechanical failure. This is because of two factors. First, the volume of the corrosion product is much greater than the volume of iron lost from the inside pipe wall. Second, as corrosion product forms, the rate of corrosion diminishes. These two factors make deep pitting corrosion of the type found on pipe exteriors a lesser concern.

Much of the existing gray CI and DI pipe installed in the United States is cement mortar lined as a remedy to this problem. DI pipe installed in water systems today is furnished with a cement mortar lining unless otherwise specified by the purchaser.

Erosion Erosion is not a typical problem in pumped mains, as velocities are normally kept in a 2–10 ft/s range per traditional design practice. Pipe size and design flow velocity are generally based on sustainable energy efficiency in pumping and minimization of surge. The maximum recommended flow velocity for cement mortar–lined DI pipe is otherwise dependent on abrasion. Parameters involved in the abrasion phenomenon include flow velocity; the amount of solid particles; the size, shape, and hardness of the particles; the type of flow (turbulent or laminar);

surface roughness and hardness of the lining; and the number of fittings per mile. Although the relative influence of these factors can be reasonably appreciated, there is no known equation able to predict abrasion resistance of different pipe materials in various situations. Inevitably, abrasion will occur at locations of changes in direction or substantial flow constrictions before it will occur along the length of a pipe barrel. Abrasion issues are generally not a problem with velocities below 15 ft/s. The abrasive characteristics of potable water are slight, although this type of water contains limited amounts of solids at much lesser flow velocity. Cement mortar–lined pipes used for drinking water service for more than 77 years have been examined and show no evidence of internal abrasion.

There are a few locations that may have extremely soft waters that can attack cement mortar linings. In these cases, calcium hydroxide, $CA(OH)_2$, can be leached out of the cement. These waters will also attack calcium silicate hydrates, which form the larger portion of cement hydrates. Although calcium silicate hydrates are almost insoluble, soft waters can progressively hydrolyze them into silica gels, resulting in a soft surface with reduced mechanical strength. Although such soft linings have still been known to be effective physical barriers for minimizing tuberculation and reasonably maintaining flow coefficients in some aggressive services, they would be more subject to abrasion phenomena, e.g., in high flow velocity or particle abrasion. In those few locations with very soft water, a seal coat on the cement mortar lining will typically retard the attack. Cement mortar–lined DI pipe is generally considered to be suitable for continuous use at pH between 4 and 12, for seal-coated linings, and between 6 and 12, for nonseal-coated linings.

Structural Damage The internal corrosion previously discussed typically does very little in the way of true structural damage. Estimates have been made that a ratio of metal consumed to tubercle size is on the order of 1 to 10. Tuberculation, therefore, will result in hydraulic failure of the pipe long before the pipe will fail structurally.

Some types of external damage, such as mishandling of the pipe during installation, might cause damage to the lining of iron pipe. For example, if a pipe were to be dropped or damaged, spalling of the internal lining could occur, with the result being the possibility of internal corrosion in that localized area. For this reason the American Water Works Association (AWWA) C104 standard and manufacturers' literature provide simple procedures for field repair of such damage.

Leakage As previously discussed, leakage caused by internal corrosion issues is unlikely to occur. As internal corrosion will not typically progress to a point of failure, leakage not attributed to external causes would most likely be associated with erosion due to leakage at a poorly installed joint or from the degradation of an older joint type. For example, in gray CI pipe systems with older joint types, soil movement and stresses over time might loosen lead-caulked joints and progress to a joint leak. The aforementioned erosion can be heightened as a high velocity stream of leaking water picks up and swirls soil particles in the path of the leak on the outside of the pipe.

4.2.2 External Surfaces

Soil Corrosion Both gray CI pipe and DI pipe require external corrosion protection in certain soil environments. Examples include soils with low resistivities and/or anaerobic bacteria, especially when compounded by significant differences in soil composition or aeration surrounding the pipe. Dissimilar metals and external stray direct currents may also necessitate additional corrosion protection. Soils contaminated by coal and other mine wastes, cinders, refuse, or salts also are generally considered corrosive. Coal and cinders contain carbon, which is cathodic to iron; mine wastes are often acidic; salts are highly ionic and conducive to the conduct of electricity; and refuse, such as in a landfill, provides a widely varied and unpredictable environment. Also considered corrosive are certain naturally occurring environments, such as swamps, peat bogs, expansive clays, and alkali soils. Soils in wet, low-lying areas are generally considered more corrosive than those in well-drained areas, and soils with the smallest particle size, such as very fine clays, are generally more of a problem than silts or sands.

Both graphitization and pitting corrosion can occur with iron pipe. Graphitization occurs when the iron constituent of the pipe is converted to corrosion products, leaving the graphite matrix of the pipe intact. Graphitization of the pipe is not readily visible, but it reduces the strength of the pipe to the point that loadings (e.g., water hammer, heavy surface load) acting on the pipe can lead to a mechanical failure. Pitting corrosion, a visible penetration of the metal, can also occur; however with iron pipe, it tends to be spread over a broad area of the pipe (generalized) instead of being concentrated at a single location.

Soil Stress Loading Soils that surround pipe tend to shift with time, coinciding with changes in moisture and surface and frost loads. This happens gradually with moderate loading, but large loads can result from expansive soil movements. Actual soil slippage occurs on slopes. Earthquakes create shock and pressure, permanent ground displacement from ground rupture, liquefaction and lateral spreading, and landslide. Each of these factors will increase loads on the pipe as well as change the pipe's bedding condition, possibly resulting in joint separation.

Third-Party Damage Third-party damage is generally accepted to be any damage that is caused to the pipe by an outside source after the pipe has been installed. Third-party damage can lead to either immediate failure of an iron pipeline or failure at a later date, depending on the severity and type of damage. This may cause internal or external issues with the pipe. Immediate damage can occur in the form of punctures or cracked pipe. Hidden damage can be in the form of damage done to internal linings that manifests itself later. Damage can also occur to an external corrosion control system that initiates a future corrosion failure. Regardless, the toughness of DI pipe is an advantage when it comes to damage resistance in general.

Another potential source of third-party damage can come from close proximity to a source of stray direct current. Although there are several ways for this to occur, the most prevalent source is from impressed current cathodic protection (CP) anode beds that are typically found in oil and gas pipelines. The magnitude of the problem is a function of the output of the anode bed, but the most important criterion is the distance, at the closest point, from the anode bed to the iron pipeline. Because impressed current anode beds are typically very far apart—perhaps miles—this is relatively rare. Failures from such stray current are going to occur very quickly, typically within one to three years.

Installation Damage Installation damage can be of the same types described previously in the "Third-Party Damage" section. Additionally, installation damage could include items such as contamination of the inside of the pipe or improper assembly of the joints. Contamination of the pipe can be avoided through sound installation practices, and cleanliness is assured by sound flushing, disinfection, and testing processes. Improper joint assembly should typically be detected and repaired during pressure testing before the pipeline is put into service.

4.2.3 Joints

Problems have resulted from the lead substitute sulfur compound joints (leadite) mentioned previously and can be related to disintegration of the jointing material, e.g., corrosion of the spigot or bell, cracking of the bell, and beam breaks in smaller diameter gray CI pipe. Problems that have been experienced are generally related to tight, moist clay soils or soils in which there are indications of the presence of sulfate-reducing bacteria. The introduction of a food or energy source (sulfur compound) together with iron in clayey soil, even with initial noncorrosive characteristics, can result in corrosion from sulfate-reducing bacteria. When joint leakage, under pressure, begins in a confined area, it can generate a tremendous erosive force that will typically lead to a more substantial failure of the joint or pipe.

The two most prominent joints for DI pipelines and late-generation CI pipelines are either the push-on or mechanical joints. These two joints are compression ring gasketed joints that work through compression of an elastomeric gasket. Of the two, the push-on joint is the most prominent because of its ease of installation and reliability. The mechanical joint is typically used in fittings but is an older technology that requires tightening bolts and is, therefore, more labor intensive. Problems with these joints are likely to occur due to improper installation. Improperly installed push-on joints are usually found during the pressure tests done prior to commissioning of the pipeline. Mechanical joints, however, if improperly installed, may not be noticed for one or two years. This is because it is possible to compress the gasket without having first properly caulked it into the recess available to it. Then, over time, the elastomer may creep into the available space, resulting in a loss of compression. When this happens, it is evidenced by a loss of torque on the nuts and bolts that are a part of the joint.

Another issue that is typically discovered during pressure testing is improper design or installation of thrust restraint. It is often the case that the highest pressure a pipeline will ever experience is during the initial pressure test. If thrust restraint is not properly designed or installed, failure can be immediate. Also, if restrained joints are used, it is an operational problem if the fitting that requires restraint is exposed while the pipeline is still under pressure. It is the interaction of the soil with the restrained joint pipeline section that provides thrust restraint. Removing the soil from around a restrained joint section without first taking the internal pressure off of the pipeline is akin to removing soil from behind a thrust block.

4.2.4 Pipelines Laid on Mounds or Blocks

In the early days of pipe laying, it was not unusual that pipelines of many materials were laid up on periodic mounds of earth or even sizable timbers, such as oak blocks. The result was that the pipe would not rest completely on the trench bottom, but would span between the supports resulting in longitudinal bending stress. Such practices have been discontinued in most areas, as it was found that they have, in some cases, contributed to eventual beam breakages or leaks in pipelines due to nonuniform load and support conditions that concentrate stress and deformation. Unless the pipeline has been purposefully and suitably designed with sufficiently frequent, properly shaped and located permanent supports and adequate wall thickness to handle the applied loads that result, any temporary supports or structures should be removed from beneath pipes. Pipelines should be provided with uniform axial support.

4.3 CONDITION ASSESSMENT TECHNIQUES

Underground pipe is subjected to both external and internal loading. Soils can shift and move, trenches can experience unequal settlement, and internal pressures can increase because of water hammer or increases in operating pressures. DI pipe with modern joining systems provides a very high safety factor in resisting service loads, but older gray CI and structurally weaker pipe materials still experience purely mechanical failures. The physical properties of gray CI and DI pipe do not change with time but the loadings may. In gray CI pipe, failures can be the result of increased loadings, corrosion, or a combination of both. In DI pipe, failures are almost completely related to corrosion. Because there was an overlap of time in which both gray and DI pipe were produced, the first task is to confirm the material being evaluated, irrespective of records. Although this can be accomplished by various means, including *sounding* of samples with a trained ear, it is probably most dependable to examine a polished area of the metal (or sample of same) with a microscope to visually observe the identifying graphite structure. If it is gray CI, the failure is typically distinct: a circumferential crack or a crack running down the barrel of the pipe. It is important to know these characteristic failure modes and to be able to evaluate whether corrosion was a contributor. In DI, it is usually a hole in the pipe, or pit, and the source of corrosion is the first question to answer.

It should also not be assumed that all pipe failures are corrosion related. Maintenance crews should know how to note probable causes of failures and how to identify significant corrosion. Proper record keeping is essential on all pipe problems. These reports should include the type and age of pipe, size, depth, location, mode of failure, type of soil, whether rocks or other structures (such as timbers) were present in the trench, the location of other nearby utilities, the observable condition of the pipe, any observed corrosion, and any other factors of importance worth noting. This information should be filed so that it can be quickly recalled as needed and used in any analysis that may follow to help prioritize future work. When potential problem areas (such as areas in which corrosive soils exist, or where shifting soils, water hammer, or other causes have impacted pipeline service) are properly reported and filed (preferably in a digitized database), such a database can be very useful in evaluating the requirements for pipeline infrastructure work.

If corrosion is suspected, a soil survey along the pipeline route can help identify aggressive areas and the extent to which they exist. Areas in which pipelines have not been installed can be surveyed ahead of time to anticipate problems, but even existing areas of the system can be sampled to help build the database. It is important to obtain sufficient samples to provide a true representation of the system's environment, but these can be obtained during routine operations, such as the installation of taps, and certainly wherever repairs are made. These data can greatly facilitate future evaluations.

External visual inspections are another method of assessing the condition of a pipeline. These can be done whenever routine operations or maintenance is performed, but also initiated through excavations made along the pipeline route, especially at locations that are of concern because of repairs or customer complaints. The pipe can be visually inspected and cleaned, with water and a wire brush or some other instrument that can remove any graphitic by-product of corrosion, and the extent to which the pipe wall is found to be sound iron. Coupons can also be cut from the pipe and sandblasted to record possible pit depths on both the outside and inside diameters.

Pipelines can also be internally inspected. If access is available, full internal inspection can be accomplished using remote control cameras. In large-diameter pipelines, it may be possible to conduct physical *manned* inspections.

Nondestructive evaluation of gray CI and DI pipelines is a developing field offering some potential in the future. The technologies are in need of improvement, and practical applications are found more in large-diameter pipelines. There can be issues with proper calibration and, depending on the technology, with the fact that the pipe is lined with cement mortar. In any case, it is generally best practice (considering the many variables potentially involved) to reasonably and dependably confirm the results of any nondestructive evaluations on iron pipelines.

4.4 CONDITION DETERMINATION

Note: The following "Rehabilitation" and "Replacement" sections are provided for information purposes only. They are not a component of the condition assessment process but are integral to an asset management program, and they were therefore included for reference.

All possible alternatives for corrective measures should be considered. Some of those alternatives are to repair as needed, replace with new properly protected pipe, or rehabilitate the pipe in place. The decisions are economical and practical and depend upon several factors, among which are the applicability and cost of possible rehabilitation technologies; the condition of the pipe, its size, location, and accessibility; its importance to service areas; and its repair frequency.

4.4.1 Repair as Needed

In this approach, repairs are not regularly required and can be done as they occur. Nothing will be done except to repair the pipe as failures occur. If corrosion is a factor, it is probable that, without remedial action, leaks will increase with time and maintenance will become too burdensome to continue at some point. If it is determined that the frequency of the breaks is low, isolated, or not increasing, then repairing the pipeline as needed is a feasible alternative. However, consideration should also be given to the effect on public relations caused by disruption of customers' service, outages due to shutdowns for repair, and the extent of possible property damage.

During pipeline repair, utility crews should be directed to check the condition of the pipeline adjacent to the break. It would also be prudent to obtain representative soil samples for subsequent testing for corrosivity. If corrosion is evident on the pipe in areas adjacent to repairs, it may be prudent to apply remedial corrosion control or to replace a section of the pipeline with protected DI pipe. In any event, all exposed pipe at the break, as well as the repair clamp, should be polyethylene encased.

4.4.2 Replacement

The next alternative to consider for a corrective measure would be to replace the existing pipeline. Again, this is an economic decision based on the size of the pipeline, the years of service, and extent to which it is still adequately sized or rated for current or future flows. Additionally, the frequency of repairs and the extent to which customer service is disrupted should be considered. The initial cost of the replacement should always be weighed against the cost of rehabilitation or repairs.

Traditionally, effective external corrosion protection for corrosive soils (in accordance with ANSI/AWWA C105/A21.5 analyses) has been achieved by use of an 8-mil thick, loose fitting polyethylene encasement. After nearly three years of collaborative work, a nine-member team of engineers from Corrpro Companies, the Ductile Iron Pipe Research Association (DIPRA), and DIPRA's member companies has developed a more extensive modern risk-based model for corrosion control known as the design decision model (DDM). The DDM is a *consequence versus likelihood* model that, in specific areas, goes into more detail than previous analyses, which evaluated only the likelihood of corrosion.

4.4.3 Rehabilitate in Place

If the evaluation of a pipeline indicates that rehabilitation is the best option, there are several options that can be considered. A discussion of those options follows in the next section.

4.5 REHABILITATION PRACTICES

4.5.1 Internal Surfaces

Where existing pipe is determined to be suffering from internal problems due to the lack of a proper lining, and the structural capacity of the pipe has been determined to be sound, a feasibility study may determine that in situ lining of the pipe is the best option. Systems are available for in situ lining of iron pipe with either cement mortar or epoxy. The decision on what type of lining to apply is going to depend on the type of service (water or wastewater, gravity or pressure), whether structural linings are required, and whether the internal pipe surface can be properly cleaned and prepared for suitable adherence and performance of the lining. AWWA *Standard C602*, "Cement-Mortar Lining of Water Pipelines in Place—4 In. (100 mm) and Larger" is available as a resource.

When existing pipe is determined to be suffering from internal, external, or joint-related issues, structural remediation via slip lining or cast in place lining of the pipe with a structural liner system may also be chosen.

4.5.2 External Surfaces

In cases in which existing pipe is determined to be deteriorating in a corrosive environment, there are several alternate remedies for which relative costs and feasibilities should be compared.

The pipe is only disturbed as failures occur and repairs become necessary. This is sometimes the best course in cases in which the pipe is readily accessible and the frequency and cost of repairs are not excessive. Upon completion of the repairs, all exposed surfaces should be encased with loose polyethylene. Mechanical joint bolts should be selected from materials such as low alloy steel with chemical composition and physical properties as required by ANSI/AWWA C111/A21.11 to minimize future bolt corrosion. To protect adjacent piping that cannot be encased in polyethylene, a sacrificial galvanic anode may be attached to the pipe prior to backfilling. This is commonly called hot spot CP.

The pipe is retrofitted for application of CP by attaching sacrificial galvanic anodes (usually of magnesium). Bonding across the pipe joints is usually necessary to provide electrical continuity if an extended length of the pipeline is to be protected.

Alternately, an impressed current system with a rectifier and anode bed is sometimes more cost-effective, especially for larger diameter pipelines or in cases in which soil conditions require high current output. Impressed current systems can, however, damage other nearby underground metallic structures, and they are generally inadvisable and impractical for water distribution systems.

Application of CP requires a higher magnitude of current output for an uncoated pipeline than for a well-coated one, and it is difficult to provide the necessary electrical continuity to all follower glands, bolts, and other metallic appurtenances. Therefore, the cost to effectively apply CP to an existing pipeline can be high. Additionally, careful design and consideration of potential stray current corrosion damage to adjacent (foreign) utilities is highly recommended.

Pipe is uncovered, cleaned, and inspected. If the pipe is found to be in sound condition, it is then encased (in place) with loose polyethylene in accordance with the appropriate standard. Encasement of an existing underground pipe is obviously labor intensive, presents difficulties in establishing uniform axial support, and is rarely cost-effective.

If the existing pipe cannot be maintained, it could be abandoned or replaced with new pipe. New installations in corrosive environments should be properly protected from external corrosion.

The best solution for corrosion control of existing structures varies with the specific installation. Any method of corrosion control is subject to the possibility of failure due to error in design or installation, but a well-conceived corrosion mitigation program will have a short payback period because corrosion failures will be virtually eliminated.

REFERENCES

ANSI and AWWA (American National Standards Institute and American Water Works Association). (2010). "Polyethylene encasement for ductile-iron pipe systems." *ANSI/AWWA C105/A21.5*, Washington, DC.

ANSI and AWWA. (2017). "Rubber-gasket joints for ductile-iron pressure pipe and fittings." *ANSI/AWWA C111/A21.11*, Denver.

AWWA (American Water Works Association). (2008). "Cement-mortar lining for ductile-iron pipe and fittings for water." *C104–08*, Denver.

AWWA. (2011). "Cement-mortar lining of water pipelines in place—4 in. (100 mm) and larger." *C602–11*, Denver.

CHAPTER 5
PIPE MATERIALS: CONCRETE

Several types of concrete pressure pipe are manufactured and used in North America. These pipe materials include prestressed concrete cylinder (PCCP), reinforced concrete cylinder (RCCP), reinforced noncylinder (RCNP), and bar wrapped concrete cylinder (BWCP). There are also a few major pipelines composed of prestressed concrete noncylinder pipe operating in the United States, and many such pipelines in other countries. Concrete pressure pipe is typically used for water transmission and distribution mains, water and wastewater treatment plants, sewer force mains, dam outfalls, and circulating water piping in power plants.

Concrete pressure pipe is an engineered product that combines the features of portland cement concrete and steel to create a structure for conveying liquids within a wide range of external loads and internal pressures.

Condition assessment of concrete pressure pipe varies depending on the type of pipe. Methods used vary from visual inspection to sounding, remote field eddy current technology, leak detection, acoustic fiber optic monitoring, impact echo assessment, and others. The condition assessment of the pipe can be performed while the pipeline is in or out of service, depending on the method selected.

5.1 HISTORY

Of these types of concrete pressure pipe, the most common are PCCP [American Water Works Association (AWWA) C301] and bar wrapped steel cylinder pipe (AWWA C303). C303 has also been referred to as pretensioned pipe, concrete cylinder pipe, or bar wrapped pipe. The following is a brief description of the various types of concrete pipe.

5.1.1 Types of Concrete Pipe

Reinforced Concrete Pressure Pipe, Steel Cylinder Type First manufactured in 1919, reinforced concrete cylinder pipe (AWWA C300), consists of a welded steel cylinder with steel joint rings at each end, a cage (or cages) of circumferential steel reinforcing bars or wire, and a structural concrete wall. Pipe diameters of 30 to 144 in. are common; however, larger diameters have also been produced (Fig. 5-1).

Prestressed Concrete Pipe, Steel Cylinder Type There are two types of prestressed concrete cylinder pipe (AWWA C301), as follows.

Lined Cylinder Pipe First used in North America in 1942, lined cylinder pipe (LCP; Fig. 5-2) consists of a welded steel cylinder with steel joint rings at each end, cast or spun internal concrete lining, high-tensile strength steel wire wrapped directly on the steel cylinder, and an external coating of portland cement mortar. Size ranges from diameters of 16 to 60 in., dependent upon the manufacturer. Typical length of pipe varies from 16 to 24 ft.

Embedded Cylinder Pipe First installed in North America in 1953, embedded cylinder pipe (ECP; Fig. 5-3) consists of a welded steel cylinder with steel joint rings at each end embedded in a concrete core, high-tensile strength steel wire wrapped on the exterior concrete core surface, and an external coating of portland cement mortar. Historically, ECP with a cast concrete coating has also been produced and continues in

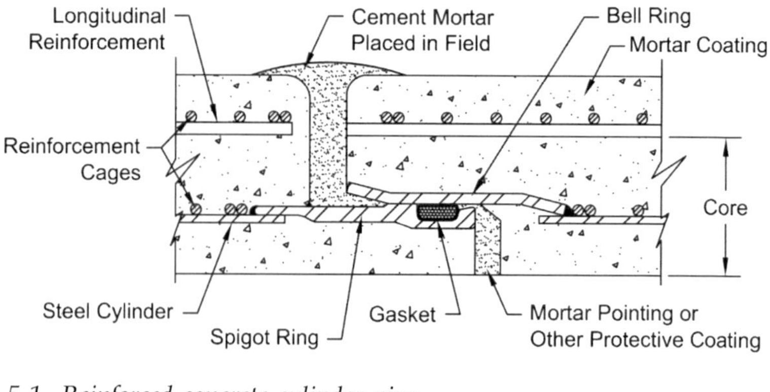

Fig. 5-1. Reinforced concrete cylinder pipe
Source: Courtesy of Forterra Pressure Pipe

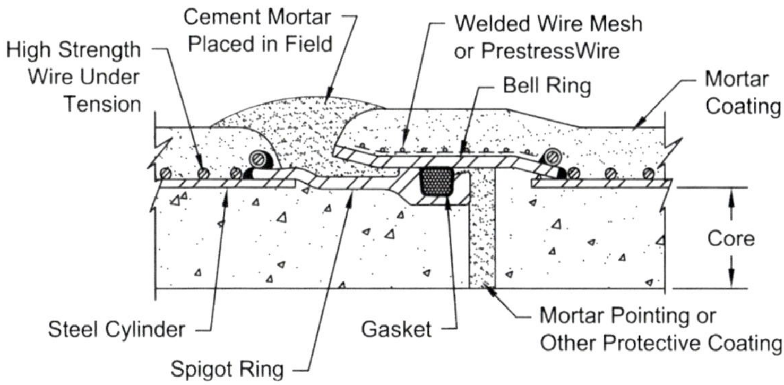

Fig. 5-2. Lined cylinder pipe
Source: Courtesy of Forterra Pressure Pipe

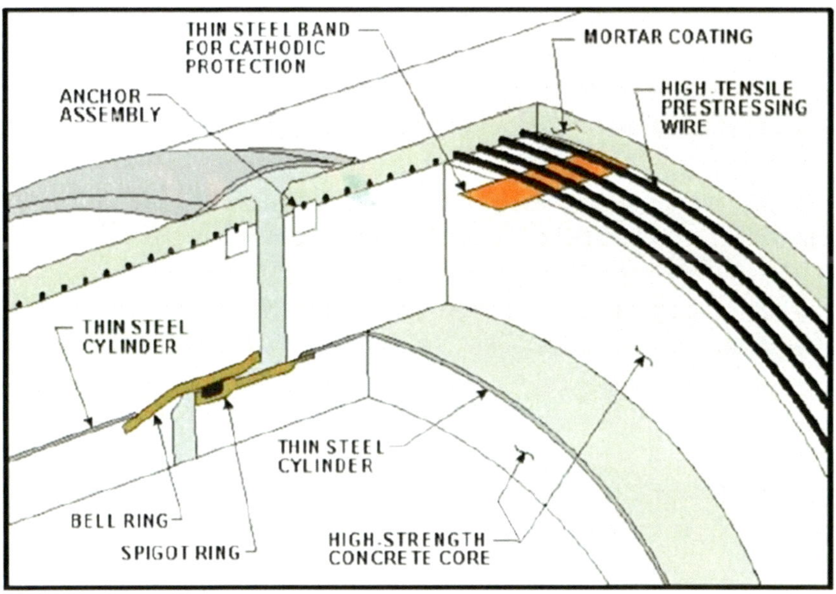

Fig. 5-3. Embedded cylinder pipe
Source: Courtesy of John J. Galleher Jr., P.E.

service. ECP size ranges typically from 36 to 144 in. and has been made in diameters of up to 252 in. Typical lengths vary from 12 to 24 ft.

Note the thin steel band in the ECP detail (Fig. 5-3). This band is commonly called a *shorting strap* and will distribute cathodic protection current, when used, to the prestressing wire without incurring significant voltage loss due to the length and small cross section of the wire. A large majority of ECP manufactured before 1990 was not made with shorting straps. Any evaluation of older ECP pipelines in which application of cathodic protection is considered should determine whether the existing pipe was manufactured with shorting straps or not.

Prestressed Concrete Pressure Pipe, Noncylinder Type Prestressed concrete noncylinder pressure pipe has been selectively manufactured in the United States and continues to be manufactured in other countries. This pipe is similar in cross section to ECP, except the steel cylinder is replaced with either a cage of circumferential and longitudinal steel reinforcing bars or wire, or longitudinal prestressing. Also, the rubber gasket joints may be either steel or formed concrete. This pipe does not have shorting straps, and it is typical for the various steel components in the pipe not to be electrically connected. Prestressed concrete noncylinder pressure pipe has lower material cost content than PCCP, but a higher labor time content. Where labor costs have risen relative to material prices, as in the United States, prestressed concrete noncylinder pipe has become less competitive in the marketplace for new installations.

Reinforced Concrete Pressure Pipe, Noncylinder Type First used in the early 1900s, reinforced concrete noncylinder pressure pipe (AWWA C302; Fig. 5-4), consists of a cage or cages of circumferential and longitudinal steel reinforcing bars/wire and a structural concrete wall with steel or concrete joint rings at each end. Diameters available are from 16 through 144 inches. This pipe is used for lower pressure applications, not to exceed 55 psi.

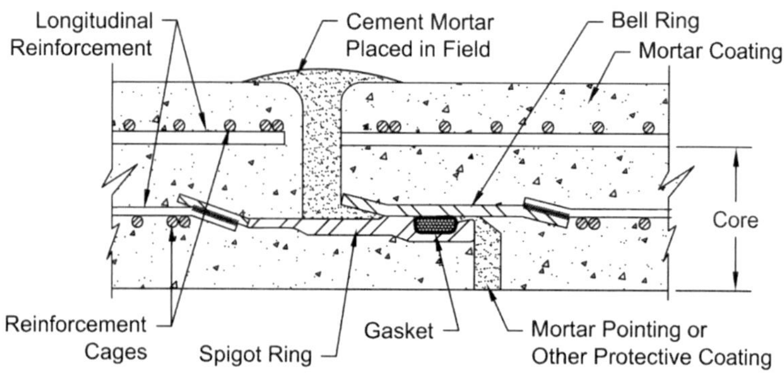

Fig. 5-4. Reinforced concrete pressure pipe, noncylinder type
Source: Courtesy of Forterra Pressure Pipe

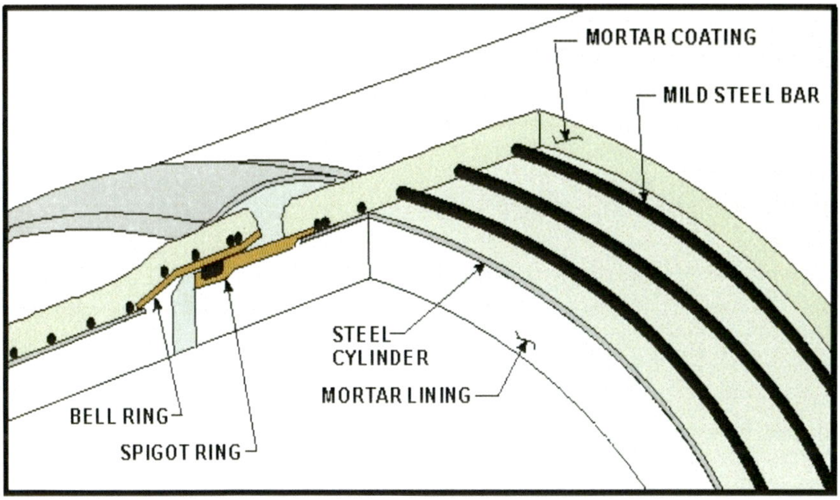

Fig. 5-5. Bar wrapped pipe
Source: Courtesy of John J. Galleher Jr., P.E.

Bar Wrapped Concrete Pressure Pipe, Steel Cylinder Type First offered in the early 1940s, bar wrapped steel cylinder pipe (AWWA C303; Fig. 5-5), consists of a welded steel cylinder with steel joint rings at each end, an internal concrete or mortar lining, spiral-wrapped reinforcing bar around the outside of the steel cylinder, and an external coating of portland cement mortar. Diameters range from 10 to 72 in. Typical length of pipe varies from 24 to 40 ft, dependent upon the manufacturer.

5.1.2 Manufacture

C300: Reinforced Concrete Cylinder Pipe The manufacture of RCCP begins with a hydrostatically tested steel cylinder that has Carnegie-style bell and spigot joint rings welded to the ends. The cylinder assembly and one or more reinforcing cages are positioned between an inside and outside form that is then cast in a vertical position with concrete.

C301: Lined Cylinder Pipe A steel cylinder is formed and attached to Carnegie-style bell and spigot joint rings. The completed assembly is hydrostatically tested. A concrete core with a minimum thickness of one-sixteenth times the pipe diameter is then placed either by the centrifugal process, radial compaction, or vertical casting. After the core is cured, the pipe core is helically wrapped with high-tensile strength wire. The wire is then embedded in thick cement slurry and coated with dense mortar.

C301: Embedded Cylinder Pipe A steel cylinder is formed and attached to a Carnegie-style bell and spigot ring. The completed assembly is hydrostatically tested. An inside and outside mold encase the steel cylinder in a vertical position, and concrete is placed inside the molds. After the core is cured, it is wrapped with a high-tensile strength wire. The wire is then embedded in thick cement slurry and coated with dense mortar. Early versions of ECP had a poured concrete coating.

Noncylinder Prestressed Pipe The concrete core for this pipe may be manufactured by casting concrete into a vertical form into which a reinforcing cage has been placed, or placing concrete into a spinning form/rack holding longitudinal prestressing wire. After the concrete is cured and any longitudinal prestressing force applied, prestressing wire is wrapped under tension around the core. The pipe is then hydrostatically tested prior to placement of the coating. A mortar or screeded concrete coating is then placed over the wire.

C302: Reinforced Concrete Noncylinder Pipe This pipe is manufactured using either steel or concrete end rings. Concrete joint rings are formed during casting of the concrete core. Steel joint rings are attached to the longitudinal reinforcement. In either case, the reinforcing cage (or cages) is placed in a vertical form and encased in the concrete core.

C303: Bar Wrapped Cylinder Pipe BWCP is designed using flexible pipe theory versus the other concrete pressure pipe types, which are designed as rigid structures. BWCP is composed of a welded steel cylinder with steel bell and spigot joint rings welded to the ends, which is hydrostatically tested. The cylinder is then lined with centrifugally placed cement mortar or concrete that is approximately ¾-in. thick. Mild steel reinforcing bar is then wrapped under nominal measured tension around the cylinder prior to coating with ¾ in. of mortar over the bar.

5.1.3 Joints

Standard There are a number of different style joints available for concrete pressure pipe. The standard is typically a steel Carnegie bell and spigot with a rubber O-ring gasket. For C302 type and noncylinder prestressed concrete pipe, a concrete rubber gasketed bell and spigot joint, or a concrete spigot with a steel bell, may be used.

Restrained Restrained joints are frequently supplied for thrust restraint requirements. They include harnessed, clamp type, bell bolt, field welded, and Snap Ring type restrained joints. These joints are not applicable to pipe with concrete spigots or bells.

Testable For concrete pressure pipe, there may be a testable joint feature available that provides for testing the joint on each section of pipe as it is installed. For this type of joint, the Carnegie spigot section is fabricated using two O-ring rubber gasket grooves. A threaded outlet between the two grooves permits pressured air testing of the seated gaskets.

Subaqueous This joint is used for underwater installations. For this application, a Carnegie bell and spigot joint with rubber gasket is generally used in combination with a special draw bolt assembly. Depending on the application, the draw bolts may or may not be left in place.

5.2 CONDITION ISSUES

5.2.1 Internal Surfaces

Corrosion Corrosion of internal steel surfaces can occur when the joints are not adequately protected or the joint mortar is damaged during installation. However, for pipelines carrying water, internal corrosion of steel joint rings becomes self-limiting such that perforation of the pipe from internal corrosion of steel joint rings on water pipelines is practically unknown. Internal corrosion of the steel cylinder can occur, owing to damage or loss of the concrete or mortar lining.

Erosion Erosion of the concrete or mortar lining can occur with high velocities or long-term wear situations such as around pumps and steep gravity flow. Generally, these problems have been associated with gravel loads in the pipe. Velocities in excess of 20 ft/s may cause erosion of the cement mortar, particularly in lower pressure pipelines in which cavitation may occur.

Aggressive Water Some attack of linings has occurred due to aggressive water that leaches calcium from the mortar or concrete. Water sources should be measured with the Langelier index or another water aggressivity index.

Structural Damage If the pipe is dropped, hit externally, or overloaded externally, a crack and/or disbondment of the mortar or concrete lining could occur. This, in turn, would provide an opportunity for the steel cylinder and joint to be exposed to corrosion. In addition, longitudinal cracking at the springline may indicate structural overloading, and longitudinal cracking at the bottom of the pipe may indicate insufficient bedding support under the haunches of the pipe.

Leaking Joints Leaking joints, especially on pipelines with chlorinated water, may allow chlorides from the interior water to become concentrated on the pipe exterior, eventually leading to corrosion of the exterior steel reinforcement of the pipe.

5.2.2 External Surfaces

Installation Damage If the external coating is damaged, it may leave the steel components (including the reinforcing steel, rod, prestressing wire, or the cylinder) exposed and without sufficient protection. Cracks in the coating may not be noticeable but hollows and disbondment can still occur. This provides an opportunity for long-term exposure to groundwater or other contaminants such as chlorides.

Third-Party Damage (Right-of-Way Management) Quite frequently, several utilities may share the same right-of-way. During routine maintenance or new construction of other utilities, water pipelines can be damaged or overloaded from construction activities. This can occur during directional drilling, installation of power poles, and general excavation. If a leak occurs immediately, the damage can be immediately located and fixed. If not, such damage could cause long-term deterioration of the pipe.

Unforeseen Aggressive Environments Installation conditions such as high chloride soils, high sulfate soils, or acidic soils can chemically attack components of concrete pressure pipe. Exposure to stray electrical currents can cause localized corrosion of concrete pressure pipe steel if the currents find the pipeline to be the path of least electrical resistance. Soils with low resistivity can be compared with a short fuse; low resistivity is not a cause of steel corrosion in concrete pressure pipe by itself, but it will accelerate corrosion started by any other mechanism or event.

The high pH of mortar and concrete forms an iron oxide layer on the surface of embedded steel that passivates the steel and prevents steel corrosion in most environments. Chlorides can migrate through the mortar or concrete and remove the iron oxide layer. If oxygen is also present, corrosion of the embedded steel will occur. Alternate wetting and drying environments will accelerate oxygenation and corrosion of the steel. Continuously wet environments will prevent replenishment of oxygen at the steel surface and essentially prevent continuing corrosion.

High sulfate soils can attack and soften concrete and mortar, which can eventually lead to exposing the steel to corrosion. Sulfate attack on mortar or concrete structures is reduced by higher cement concentration, lower tricalcium aluminate cements, low permeability of the mortar or concrete, and especially a lack of a sulfate concentration mechanism (i.e., wetting and drying or evaporation from a partially buried structure). Damaging sulfate attack of concrete pressure pipe is uncommon, and locations where such attack has occurred are typically known to local pipeline designers or operators.

Acid soils, damaging to concrete pressure pipe, are usually associated with artificial concentrations of contaminants, such as mining waste or industrial dumps. In such locations, the pipe should be isolated from the acidic soils by impermeable envelopes of plastic, high-build coatings, or clay.

Stray electrical currents from electric-powered transportation systems, cathodic protection systems on other structures, or similar sources may find a concrete pressure pipeline to be the path of least electrical resistance. If this occurs, corrosion of the concrete pressure pipe steel will occur where the current leaves the pipe.

Deficient Prestressing Wire and/or Coating From approximately 1973 to 1984, AWWA C301 prestressed pipe was allowed to be made with prestressing wire that had no maximum tensile strength limit. The stronger wire, often referred to as Class IV wire, was frequently dynamically strain aged during drawing, resulting in less ductility and increased susceptibility to damage from hydrogen embrittlement. Hydrogen can be generated either by excessive cathodic protection currents or from ordinary corrosion at the bottom of pits or cracks in such wire. Any resulting embrittlement could cause this wire to break under normal design stress. In addition to reinstating upper tensile limits for the wire, a torsion test was added to the ASTM International standard A648 wire specification in 1988 to better ensure new wire was not dynamically strain aged.

Also during the 1970s to early 1980s, many mortar coatings were applied with low moisture contents in the mix. This resulted in coatings with high permeability, increasing the probability of corrodents penetrating the coating and attacking the steel wire or bar.

Experience has shown that dynamically strain-aged wire protected by a lower permeability mortar coating in an environment that does not initiate corrosion can provide good pipeline performance.

5.2.3 General

With the exception of lining erosion from gravel loads in the pipe, there have been essentially no structural failures of C300 and C302 pipes. PCCP (C301) experienced a number of failures due to poor quality coatings, wire susceptible to hydrogen attack, and unanticipated chloride penetration of the mortar coating in areas with cyclic wetting and drying. These issues have been addressed in the AWWA C301 standard and the AWWA

Manual M9, *Concrete Pressure Pipe*, but much pipe with these issues remains in service. Failures of C303 pipe have been generally associated with localized corrosion and improper placement of bedding and backfill.

5.3 INSPECTION PROCEDURES

Inspection procedures and condition assessment of PCCP are described in more detail in the 2012 Water Research Foundation and EPA publication "Best Practices Manual for Prestressed Concrete Cylinder Pipe Condition Assessment: What Works? What Doesn't? What's Next?" Selection of appropriate inspection technologies depends on a number of factors and may be performed in a cost-effective way by first performing an engineering evaluation of the pipeline or the pipeline system to determine pipeline criticality as described in the manual. Some of the available technologies include the following.

Environmental Review General observation of the installation conditions of the pipeline will provide clues regarding areas of potential pipe degradation. Identification of soil resistivities, crossings of other pipelines or similar potential sources of stray electrical currents, creek crossings or irrigated areas that might cause cyclic wetting and drying, identification of abandoned waste pits or other potentially harmful environments, and location of shallow groundwater along the pipeline route can provide a sharper focus for additional inspection.

Visual Inspection and Sounding Sounding inspections rely on impacting the surface of the pipe wall and listening for hollow sounds that may be indicative of delamination and prestress loss, especially when accompanied by longitudinal cracking. Hollow sounding areas may also be caused by variations in pipe wall such as placement of steel cylinder, bedding, and other factors not indicative of significant damage. Sounding can be conducted with a steel rod or hammer, hitting the interior of the pipe for initial inspections and the exterior of the pipe for location and potential confirmation of external problems indicated by other inspection techniques. The steel rod or hammer produces a resonance frequency at the location of delamination that can be heard as a *thud* compared with the regular *ting* sound.

Remote Field Eddy Current (Electromagnetics) Electromagnetic testing detects and quantifies the number of breaks in the prestressing wire of PCCP. This type of inspection provides water and wastewater system operators with information on the location and approximate number of wire breaks along the length of the pipeline. See the "Pipe Inspection Tools" chapter for further information.

Acoustic Monitoring Acoustic monitoring can be performed with either discrete sensors along the pipeline or with an acoustic fiber optic technology. Sensors continuously monitor the acoustic activity in a pipeline to identify the acoustic events associated with the failure of prestressing wires and report the time and location of breaking prestressing wires in each pipe. This information can then be fed into a structural model to assist the pipeline operator with determining when the pipeline needs to be repaired, using the actual rate of deterioration. See the "Pipe Inspection Tools" chapter for further information.

Impact Echo Technology Impact echo is a method for nondestructive evaluation of concrete and masonry, based on the use of impact-generated stress (sound) waves that propagate through the structure and are reflected by internal flaws and external surfaces. Impact echo can be used to determine the location and extent of flaws such as cracks, delaminations, voids, honeycombing, and debonding of the pipe core. This disbondment may indicate that prestressing wires, if present, have broken. See the "Pipe Inspection Tools" chapter for further information.

Leak Detection Leak detection, using internal and external sensor technologies, is used to identify joint leakage as well as leakage through pipe wall in pipes in an advanced state of distress.

Developing Technologies There are technologies under development to assess pipeline condition, including pipeline average wall stiffness technology and others described in the Water Environment Research Foundation (WERF) report, "Condition Assessment for Drinking Water Pipelines" (2013).

5.4 CONDITION ASSESSMENT

The pipeline operator needs to look at all information that is available on the pipeline such as design, manufacturing, installation, operation, hydraulics, maintenance, and right-of-way encroachments to determine both the likelihood and consequence of a pipeline failure (see Chapter 3—Planning). These data are used in an asset management and pipeline condition assessment approach to determine pipeline criticality and select an appropriate inspection technology.

Risk Curves Technology Risk curves technology (RCT) is a method for determining the failure risk of an individual distressed PCCP and prioritization of pipe repairs given the size of distress and the maximum working and transient pressures. Margin to failure is determined using the failure risk curves corresponding to limit states related to serviceability, damage, and ultimate strength of the pipe with broken wires. Repair priorities are assigned to pipes with broken wires to identify pipes with unacceptable margin to failure when subjected to the maximum internal pressure and gravity loads (Zarghamee and Ojdrovic 2001; Zarghamee et al. 2003). The research done in support of the RCT included hydrostatic pressure testing of PCCP with broken wires to failure, field inspection of pipes with broken wires, nonlinear finite element analyses to further substantiate and calibrate the model, and development of a procedure to construct the risk curves. RCT helps operators manage the pipelines and maintain failure risk and repair budgets at an acceptable level by identifying pipe pieces that need to be repaired immediately, in five years, or just need to be monitored and repaired later during scheduled outages. RCT can also be used to help identify an appropriate inspection technology by determining the level of damage necessary for pipe failure at the maximum expected pressure.

5.5 REHABILITATION PRACTICES

Note: The following rehabilitation and replacement sections are provided for information purposes only. They are not a component of the condition assessment process but are integral to an asset management program, and they were, therefore, included for reference.

Closure and Repair Assemblies If an individual pipe needs replacement because of contractor damage or failure, a closure assembly can be installed. A typical closure assembly consists of two fabricated short pipes complete with one plain steel end on each and either using a bolted sleeve coupling or welding a split sleeve butt strap to seal the closure.

Relining For larger diameter pipe (>48 in. or 1,200 mm), placement of a steel lining may be used to rehabilitate long stretches of the pipeline. This is accomplished by fabricating steel sections, generally in lengths equal to the lengths of the original pipes. The liners are split longitudinally to allow them to be partially collapsed like a clock spring during insertion, then expanded back into a circular cylinder. The longitudinal and circumferential seams are then butt welded to provide closure. The annular space between the steel reline and the inside diameter of the host pipe is grouted with a cementitious material. The inside of the steel liner is typically cement mortar lined. The finished steel reline may be up to 6 in. (150 mm) in diameter smaller than the original pipe.

Plastic liners such as high-density polyethylene have also been used to line pipe attacked by erosion from gravel loads and other impingement.

To perform the construction, the original pipeline is taken out of service and dewatered. An access and insertion pit is constructed to provide access for the installation. Fittings and outlets may be accommodated either by transitioning to the existing fittings, if the condition of the fittings is not in question, or building the new liner to match existing pipeline deflections and tie to the outlet piping.

External Posttensioning PCCP can be repaired using an external posttensioning strengthening system. This repair is accomplished by wrapping the pipe with tendons and posttensioning the tendons to the desired strength requirements. The posttensioning is typically protected by fiber-reinforced shotcrete. The pipe must be excavated so that the external system can be applied. For this application, the pipeline does not have to be taken out service; however, a reduction in pressure may be required during the excavation and construction. This application is appropriate for spot repairs.

External Steel Cylinder Similar to external posttensioning, this repair process is an external repair that does not require the pipe to be taken out of service. The pipe must be excavated so that two halves of an external cylinder can be placed around the pipe with longitudinal seams butt welded. The cylinder is placed with an annular space surrounding the pipe and sealed on the ends so the annular space can be filled with nonshrink grout. The grout allows for direct stress transfer from the pipe's steel cylinder to the external steel cylinder by compression of the pipe concrete and mortar and the nonshrink grout. The outside of the external steel cylinder is then protected with field-applied mortar.

Dig and Replace For replacing several consecutive sections of pipe it may be necessary to remove the existing pipe and provide for complete replacement. This can be accomplished by installing new pipe pieces and a closure assembly with a bell and spigot adapter.

Carbon Fiber–Reinforced Polymer PCCP strengthening, using internally and externally bonded carbon fiber–reinforced polymer (CFRP) liners, has been successfully used since the mid-1990s. Carbon-based fiber-reinforced composites are high strength, noncorrosive, and durable materials that can add significant strength. This makes them very suitable for PCCP strengthening. Carbon fiber–reinforced liners are constructed on site using multiple layers (plies) of carbon fiber saturated with epoxy adhesive and bonded to the inner or outer surface of the pipe manually (hand lay up). The thickness of the CFRP composite liner varies from 0.25 to 0.75 in., depending on pipe diameter, internal pressure, and design approach. For internal repairs, the pipe does not need to be excavated, and repair can be performed through existing manholes.

REFERENCES

ASTM. (2012). "Standard specification for steel wire, hard drawn for prestressing concrete pipe." *ASTM A648*, West Conshohocken, PA.

AWWA (American Water Works Association). "Concrete pressure pipe." *AWWA M9*, Denver.

AWWA. "Concrete pressure pipe, bar-wrapped, steel-cylinder type." *AWWA C303*, Denver.

AWWA. "Prestressed concrete pressure pipe, steel-cylinder type." *AWWA C301*, Denver.

AWWA. "Reinforced concrete pressure pipe, noncylinder type." *AWWA C302*, Denver.

AWWA. "Reinforced concrete pressure pipe, steel-cylinder type." *AWWA C300*, Denver.

Water Research Foundation and EPA. (2012). *Best practices manual for prestressed concrete cylinder pipe condition assessment: What works? What doesn't? What's next?*, Denver.

WERF (Water Environment Research Foundation). (2013). *Condition assessment for drinking water pipelines*, Alexandria, VA.

Zarghamee, M. S., Eggers, D. W., Ojdrovic, R. P., and Rose, B. (2003). "Risk analysis of prestressed concrete cylinder pipe with broken wires." *Proc., Pipelines 2003: New Pipeline Technologies, Security, and Safety*, ASCE, Reston, VA.

Zarghamee, M. S., and Ojdrovic, R. P. (2001). "Risk assessment and repair priority of PCCP with broken wires." *Proc., Pipelines 2001: Advances in Pipelines Engineering and Construction*, ASCE, Reston, VA.

CHAPTER 6

PIPE MATERIALS: STEEL

Steel pipe is commonly found in both aboveground and below ground installations and is generally of a large enough diameter to allow for internal visual inspection. This large size is due to the primary use for steel pipe in the transmission of water rather than for distribution purposes. Modern steel water pipe will, in nearly all cases, have some sort of protective lining and coating. Older installations may lack lining or coatings, making it more susceptible to attack from corrosion.

The most common concern in assessing the condition of steel pipe is the amount of corrosion that has occurred in the steel wall. Pitting is the general form of corrosion that occurs. Consequently, steel pipe rarely exhibits either structural collapse or separation due to mechanical forces. Rather, the normal failure mode of corroded steel pipe is localized leaking.

For buried steel pipe, the primary design consideration is circumferential resistance to the internal pressure. Leakage occurs when the localized area of the steel shell is reduced to a thickness that creates a local stress greater than that of the tensile strength of the steel. Leakage may also occur at weakened joints or riveted seams found in older pipelines. For conservative design, the steel shell stress is typically limited to 50% of the yield strength for working or operating pressure and 75% for transient pressures. As steel pipe is frequently lined, coated, and buried, determining any localized shell reduction may be difficult.

Exposed pipe offers advantages of evaluation that buried pipe effectively precludes: specifically, visual inspection of the exterior surface. Internal visual inspections can be completed on larger diameter lines when they are taken out of service. Another option (internal smart pigging of steel pipe) can be done, but it has two potential drawbacks: the diameter of the pipe may be too large for any commercially available pig to be effectively used; also, the presence of cement mortar or coal tar enamel lining prohibits an accurate reading of the steel shell. Ultrasonic inspection is available to evaluate the exterior of an exposed pipe as these lines usually have a relatively thin film coating. Also, evidence of localized deteriorated exterior coating provides the appropriate location for an expanded investigation. Although exposed steel pipe is designed for internal pressure much like buried pipe, it must also be capable of withstanding any vacuum condition and structural shear and bending forces when installed on piers or saddles. Determination of the ability to resist these latter forces is generally associated with a visual inspection of the exposed pipeline.

6.1 HISTORY

Steel pipe was first produced in the United States in the late 1850s, shortly after the Bessemer process for making steel was invented in 1855. With the advent of the open-hearth process in 1861, steel became even more economical and available. Steel pipe provided designers with the ability to greatly reduce the amount of metal required in a pipeline. However, due to a lack of knowledge as to the nature of the pipe and soil interaction for an economical flexible pipe design, many steel pipelines were designed using a rule of thumb. Wall thicknesses found in old pipe from this era reflect this approach. Some value for a corrosion allowance was also somewhat common and added to the rule-of-thumb design thickness approach. Diameters were not standardized, as steel pipe had the ability to be cold formed into a cylinder of any dimension and joined with methods such as rivets or lock bars (Fig. 6-1). From available records, many of these 100-year-old (or older) pipelines are still in service.

Fig. 6-1. Riveted steel pipe (left) and lock-bar steel pipe (right)
Source: (left) Courtesy of Historical Photo Collection of the Department of Water and Power, City of Los Angeles; (right) Courtesy of Northwest Pipe Company

6.1.1 Cylinder Manufacture

The first steel pipelines were rolled sheets with riveted seams. Joints were made by crimping one end of a pipe and stabbing into the non-crimped end of the next, much like stove pipe. The first known installation of steel pipe was in 1858. Until 1890, existing records indicate more than 2 million ft of steel pipe was installed in the United States.

In 1905, a new steel pipe making process was developed in which two half troughs were formed and joined to form a circular section using two clamps (with an H-shaped cross section) called lock bars. This method was very popular until approximately 1930, during which time approximately 3.3 million ft of lock bar pipe was installed in the United States versus approximately 1.5 million ft of riveted seam pipe.

Electric welding began as a novelty in the 1920s and became very popular in the 1930s when welding machines and fluxes became common and readily available. From 1920 to 1940, approximately 7 million ft of welded steel pipe was installed in the United States. During World War II, almost all steel production was employed by the war effort, and there is little record of steel pipe usage. Following the war, spiral fusion welding started becoming available and pipelines began increasing in size and scope.

6.1.2 Linings

Steel pipe has generally required a lining to protect the interior steel surface from corrosion. A wide array of linings has been employed. Early pipes were largely unlined; although meager coatings of bitumen, asphalt, mineral rubber, or galvanizing were used. Until the late 1920s, the dominant protection method was hot dipping the pipe in asphalt or asphalt derivatives. Around that time, coal tar enamel became available, initially applied by the hot dipping method and, ultimately, lined by centrifugal spinning. All of these methods provide a physical barrier to the steel surface to prevent corrosion. Cement mortar lining was first employed in the late nineteenth century, but it was not until the 1920s that a practical method of in-plant application, using a cast and pulled plug method, was introduced. Cement mortar prevents corrosion by creating an alkaline environment at the steel surface, passivating the surface to prevent corrosion. Despite the improved in-plant application process, cement mortar lining did not become popular until the 1950s. Coal tar enamel lining of steel water pipes continued in popularity until the 1970s. Whereas in-place cement mortar lining had been used prior, it was in the 1930s that the modern-day, practical, and economical process was first employed. Epoxy linings, initially with a coal tar base, experienced limited application, primarily from the late 1970s. Beginning in the 1990s, polyurethane linings began being used. Today, cement mortar lining is by far the most commonly applied protection for the interior of steel water pipes.

6.1.3 Coatings

Much like linings, early steel pipe was generally uncoated or coated primarily with a hot dipping of asphalt or asphalt derivatives. Coal tar enamel became popular in the 1920s, initially by the hot dipping method of application and later by spray application of the slowly rotating pipe. Later, this coating was reinforced by the addition of an intermediate wrap of asphaltic felt or fiberglass fabric. To facilitate handling considerations, coal tar enamel coating received a final application of white wash or Kraft paper. In the late 1970s, tape wrap coatings, consisting of two to three layers (80 mils) of polyethylene tape with a butyl rubber adhesive, began being routinely used. Soon thereafter, polyurethane coatings were introduced and began gaining in popularity. Today, tape coating is the most popular dielectric coating followed closely by polyurethane. Also, in parts of the United States (generally in the far western states), cement mortar coating has been successfully used since the 1950s on steel pipelines placed in less aggressive soils. There have also been applications of cement mortar being applied over both coal tar enamel and tape wrap coatings to act as an armor coat or rock shield, providing added mechanical protection. This latter practice was also largely relegated to the far western states.

6.1.4 Field Joints

As mentioned previously, very early joints were crimped and driven home much in the manner of stove pipe joints. These were followed closely by riveted, flanged, and bolted sleeve type joints. Fig. 6-2 shows a riveted joint on a welded steel pipe and a bolted type coupled joint on a welded

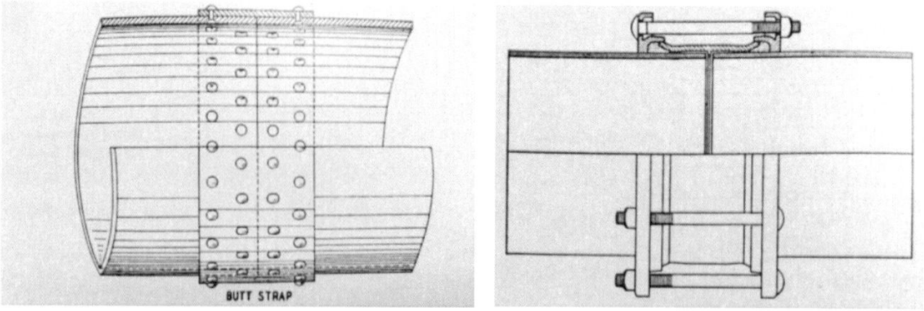

Fig. 6-2. Riveted joint (left) and bolted type coupled joint (right) on welded steel pipe, circa 1940s
Source: ARMCO Drainage and Metal Products, Inc., Handbook of Welded Steel Pipe, *1950*

steel pipe, respectively, from the 1940s. Welded joints gained popularity in the 1930s, with lap-welded joints being the primary application for buried pipelines and butt-welded joints used to resist flexural or shear stresses on exposed pipelines.

Gasketed joints started becoming popular for buried pipelines in the 1950s, initially in diameters up to 36 in. and eventually to diameters up to 72 in. or greater. These gasketed joints generally consisted of a flared bell end and either a rolled groove or Carnegie-style spigot. A cross section of a typical modern-day rolled groove joint is shown in Fig. 6-3. Today, virtually all buried steel water pipelines are constructed of either a combination of gasketed joints (for nonthrust portions) and lap-welded joints (to resist differential forces such as thrust, seismic, or landslide) or are constructed entirely with lap-welded joints. Limited cases of butt-welded joints may be found in geohazard areas such as fault zone crossings or areas prone to soil subsidence. They may also be found in long span areal crossings and some tunnel installations.

6.2 CONDITION ISSUES

6.2.1 Internal Surfaces

Corrosion Corrosion takes the form of a localized pitting attack. This occurs when the lining has deteriorated to the point that bare metal is exposed or in cases in which the steel pipe is installed unlined. Corrosion of

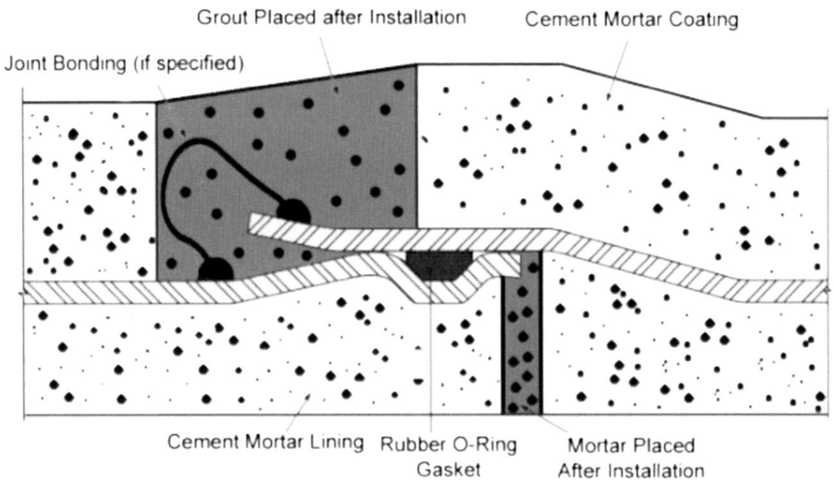

Fig. 6-3. Rolled groove gasket joint
Source: Courtesy of Northwest Pipe Company

the pipe wall can also occur at joints when field joint lining material is applied improperly. If the remaining wall thickness is less than that of the allowable internal pressure design, the resultant condition is not as conservative as the basic pipe design. At the point of attack, as the stress in the remaining wall thickness approaches the tensile strength, the pit is in danger of leaking.

Erosion Erosion occurs with higher velocity or long-term wear situations. Generally, velocities in excess of 20 ft/s may cause erosion of a cement mortar lining. Although lesser flows may have the same effect if high loads of abrasives flow through the pipe or in highly aggressive waters that leach calcium hydroxide from the mortar. All of these conditions are extremely rare. Over time, thin film linings, such as epoxies, polyurethanes, or coal tar enamel linings, can also exhibit wear down to bare metal due to erosion.

Structural Damage Although the cause of structural damage normally does not occur on the interior surface of the pipe, external damage may be displayed in the form of an indentation in the pipe. In the case of a cement mortar–lined pipe, or possibly an enamel-lined pipe, this condition would likely result in the localized loss of lining and possible onset of internal corrosion. Polyurethane lining may conform to the indentation. It is important that the indentation not show a distinct crease. Regardless of the condition of the interior surface, the cause of the indentation and the condition of the pipe exterior coating must be investigated.

Leakage The form of wall leakage is generally pinholes in the pipe. If these pinholes are associated with internal pits, the cause is associated with an internal condition. Joints may also leak with time, particularly riveted joints that undergo some vibration. Improper installation such as damaged gaskets or incomplete or defective welds can also be a source of leakage, but these are typically discovered during the installation inspection process.

6.2.2 External Surfaces

Soil Side Corrosion External corrosion also takes the form of localized pitting. Older generations of steel pipe, particularly riveted steel pipe installed prior to the 1930s, often had limited or no coating applied, making it more susceptible to corrosion activity. For coated pipe, soil side corrosion occurs when the dielectric coating has holidays (penetrations to the metal surface) or where the coating is inadequate for the corrosive condition (aggressive soils with cement mortar coating or insufficient dielectric material application). Generally, the protection afforded external field joints has not been problematic as various tape and shrink sleeve wrapping measures for all dielectric coatings and cement mortar grout band applications have proven a much simpler procedure than the protective mechanisms required for internal joint protection.

Stray Current Stray current is caused by stray DC current collecting on the pipe wall. If not addressed, it can cause rapid metal loss at the location where the current leaves the pipe. Some common sources of stray current are neighboring pipelines with impressed cathodic protection (CP) systems and DC powered transit systems such as light rail.

Soil Stress Loading All soils creep with time. This action may overstress the coating, particularly in the case of thin film coatings or nonreinforced coal tar enamel. This overstressing will cause failure of the coating and probable corrosion.

Third-Party Damage Excavation required for new construction may damage an existing pipeline. Generally, in the case of steel water lines, such damage takes the form of cut or stripped coating and possible pipe wall indentation. It does not necessarily involve an immediate leak or failure. However, direct impact by horizontal drilling installations may lead to an immediate failure and repair.

Installation Damage Although any installation should be professionally inspected, particularly in the case of large-diameter transmission pipelines, this is unfortunately not always accomplished. Damaged coating is the primary concern. Equally, any wall indentations observed should have been thoroughly investigated. Any wall penetration, rolled joint gasket, or incomplete joint weld should be detected during the field hydrostatic test. In an operating pipeline, damaged coating can be identified using a close interval survey procedure. Wall penetration and leaking gasketed or welded joints can be detected by one of the leak detection methods.

6.3 INSPECTION PROCEDURES

There are two readily recognized conditions that warrant further inspection of a steel pipeline: leakage and general water loss. Leakage may be recognized by surface detection, whereas general water loss usually relates to a number of small leaks and is determined with a system investigation by competent water resource professionals.

Various techniques are available for investigation and inspection, some of the most common for steel pipe are found next; a more detailed list of available techniques can be found in Chapter 10, "Pipe Inspection Tools."

Man Entry In larger diameter water lines, when a shutdown of the line is possible, man entry can provide valuable information as to the internal condition of the pipe. Lining deterioration is readily detected, as is pitting attack. Any indentations can be observed as well as joint or other pipe conditions that indicate differential settlement.

Ultrasonic Inspection Unlike their use in the inspection of petroleum pipelines, smart pigs are not practical for steel water pipelines. They require launching and retrieval sites that are built into petroleum lines but rarely into water lines, which are generally of relatively large diameter and cannot provide meaningful readings when the pipe has a cement mortar or thick film lining. Handheld ultrasonic inspection can augment internal man entry visual inspection and can provide information on pit depth and remaining steel thickness on steel pipe that has no coating or coating damage at the surface to be tested.

Acoustic Inspection Acoustic inspection entails sending a receiver through a pressurized pipeline and listening for telltale sounds that indicate leakage. The level of the sound can determine the size of the leak. The approximate location of the leak can also be determined.

Close Interval Survey This survey is accomplished by walking at grade above the pipeline and briefly inserting a transmitting device approximately every 2½ ft. By having an electrical connection between the pipeline and a receiver, the condition of the coated surface of the pipeline can be determined and individual holiday locations recognized.

Closed-Circuit Television Closed-circuit television can be used for small-diameter pipelines that can be dewatered adequately to permit installation of the device. It is not practical for large-diameter pipelines, as manned entry can provide more substantial information such as specific knowledge of the lining condition.

6.4 CONDITION DETERMINATION

To provide input to a rating system on the condition of a steel water pipeline, information obtained from the inspection must be analyzed both objectively and subjectively. This will be particularly true for very old pipelines that may be difficult to analyze due to poor to nonexistent coating and lining protection, seam and jointing systems that may pose future concerns, or pipelines that have exceeded their effectiveness due to higher flow needs. Some considerations would include the following:

Schedule for Future Observation This would be particularly likely if the interior protection was found to be in a state of deterioration but not of sufficient present concern to warrant immediate repair. It may also be suggested in cases in which a riveted pipeline had shown some signs of relaxation in a few seams but not an extensive concern. The cost of a present remediation would need to be compared against the cost of future investigation as well as the probability of continued or eventual substantial deterioration.

Rehabilitation Depending on the nature of the concern, rehabilitation would generally pose a likely scenario for a pipeline showing distress. As structural failure of a steel water pipeline is not a usual concern, rehabilitation and remediation measures may be economically accomplished.

Replacement This would be expected to be a last resort consideration. It would likely be high in the rating system should the pipeline be very old and in critical need of multiple repair procedures or if it no longer met the needs of the system.

6.5 REHABILITATION PRACTICES

Note: The following rehabilitation and replacement sections are provided for information purposes only. They are not a component of the condition assessment process but are integral to an asset management program, and they were, therefore, included for reference.

6.5.1 Internal Surfaces

Cement Mortar Lining For unlined or deteriorating thin film lined pipe, in situ lining with cement mortar may be an economical solution. Diameter, number and degree of direction changes, and degree of deflection (ovality) of the pipeline affect this option. Relining of field-applied cement mortar joints should be accomplished where the joint material exhibits a soft, flakey, or worn appearance.

Slip Lining This method of rehabilitation is generally associated with potential structural concerns for the pipeline. It would not be expected to be warranted for steel pipe as collapse of the pipeline should not be a concern. Extensive leakage, requiring extraordinary repair, would likely have been prevented by being addressed earlier in the life of the pipeline.

Welding Repair Welding of pinholes or deep pitting areas can be accomplished in larger diameter pipelines. Depending on the repair area, this may be accomplished by either directly welding the affected area or by welding a plate over the section requiring repair. This repair procedure would be accomplished in conjunction with spot relining with epoxy, polyurethane, or cement mortar.

Interior Joint Repair Assemblies These mechanisms are available from a number of companies specializing in repair of leaking joints.

6.5.2 Exterior Surfaces

Coating Repair or Replacement Coating repair of buried pipe may be an option where a close interval survey detected flaws in the coating or for exposed pipe with limited coating issues. Coating replacement would only be an option for exposed pipe with consideration given to the effect of any blast medium on the environment.

Welding Repair This procedure would repair pinholes or deep pits in the exterior of a steel pipe. It should be followed by coating repair or replacement, as well as lining touch up, if possible.

Clamps or Weld-on Plates For extensive third-party damage occasioned by excavating equipment or errant horizontal drilling, split sleeve type clamps may be provided to repair the damage. An inexpensive alternative, eliminating the need for full pipe excavation, would be a curved plate welded to the exterior surface of the pipeline. For either procedure, all metal surfaces should be adequately lined and coated.

CP For long-term remediation of external corrosion, either occurring or having the potential to occur on a steel pipeline, CP usually offers an effective, cost-effective corrosion remediation solution for a buried pipeline. As the function of CP is to protect areas of a steel pipeline that have holidays, the addition of CP to an existing pipeline may eliminate the need for localized coating repair. One requirement for an economical CP application is for the pipeline to be electrically continuous. This is generally the case with a welded or gasketed joint steel pipeline, as gasketed joints are usually electrically bonded.

REFERENCES

ARMCO Drainage and Metal Products, Inc. (1950). *Handbook of welded steel pipe*, Berkeley, CA.

CHAPTER 7

PIPE MATERIALS: FIBERGLASS

Fiberglass pipe is a composite structure manufactured from glass fiber reinforcements embedded in or surrounded by cured thermosetting resin. Properly the product is known as fiber-reinforced plastic (FRP) or glass-reinforced plastic (GRP), in popular usage the product type is referred to as *fiberglass*. Alternate nomenclature for fiberglass pipe is glass fiber reinforced thermosetting resin pipe (RTRP) and glass fiber reinforced polymer mortar pipe (RPMP). The RTRP is fiberglass pipe without an aggregate, and the RPMP is fiberglass pipe with an aggregate.

As with many other composite materials, the two materials act together, each overcoming the deficits of the other. Whereas the plastic resins are strong in compressive loading and relatively weak in tensile strength, the glass fibers are very strong in tension but have no strength against compression. By combining the two materials, FRP becomes a material that resists both compressive and tensile forces well.

This document describes how to inspect FRP pipe that is already in service (used) to determine its condition, knowledge of which is used to assess fitness for continued service and to plan repairs. Three complementary types of inspection are described: (1) external inspection with the equipment in service, (2) external inspection with the equipment out of service, and (3) internal inspection. In general, for all three inspections, equipment should be down to allow for a complete inspection.

7.1 HISTORY

7.1.1 Early Development

Various milestones trace the early development leading to current day FRP:

1936 Polyester resin patented by DuPont
1939 First successful commercial manufacture of glass fibers by Owens Corning
1944 Fiberglass boat hulls produced for the U.S. Navy
1948 Fiberglass pipe introduced into the oil industry
1953 Fiberglass car body produced for Chevrolet Corvette
1960s Fiberglass boat hulls gain larger market share than wood boat hulls
- First fiberglass aircraft manufactured
- First fiberglass underground tanks and pipe manufactured

1970s Military and other national and international standards [e.g., ASME, ASTM, American National Standards Institute, American Petroleum Institute, American Water Works Association (AWWA), National Sanitation Foundation, Underwriters Laboratories] introduced to standardize manufacturing, testing, and performance criteria for FRP to address specific anticipated service applications

7.1.2 Types of Pipe and Joints

What goods (pipe and joints) can be expected to be found in service by water works professionals and end-users?

The focus of this chapter is on all FRP products that could be potentially encountered in the water works distribution and transmission marketplace. There were numerous idiosyncrasies in the various manufacturing, testing, and component elements used through the period leading to the maturity of FRP found today. Therefore, the historical types of FRP goods that a water works professional may encounter in water transmission and distribution pipelines are often referenced by the manufacturers of the goods not necessarily the component materials, testing, components, or the representative manufacturing processes. Historical FRP goods no longer produced, but which can be found in the water works transmission and distribution industry, are normally identified by their trade names, examples include Techite, Flextran, and CorBan. However, FRP goods currently being manufactured for use in the water works transmission and distribution industry are usually identified by the type of manufacturing process employed, such as continuous advancing mandrel, reciprocal rotating fixed mandrel, and centrifugal casting.

The types of joints that may be encountered by the water works professional, regardless of the type of goods encountered, are

- Flanged (available in all diameters);
- Bell and spigot, adhesive-bonded (generally available up to 24 in.);
- Reinforced overlay [butt and wrap field lamination type (available in diameters up to 156 in.)];

- Integral bell and spigot, gasket seal [single or double gasket (generally available in diameters up to 60 in.)]; and
- Gasketed coupling joint (available in diameters up to 156 in.).

The pressure rating of the joint supplied and potentially encountered is equal to or greater than the pressure rating of the pipe goods.

7.2 CONDITION ISSUES

7.2.1 Internal Surfaces

Corrosion As the materials of construction of FRP are nonmetallic, corrosion in a water service application is not a consideration.

Erosion FRP is highly durable and resistant to internal wear and erosion. An abrasion resistance test is routinely performed by current manufacturers. The most common test protocol is the Darmstadt rocker test. The expected abrasion loss following the Darmstadt test protocol, using a mixture of sand, gravel, and water, is typically found to be in the range of 0.20–0.40 mm after 100,000 cycles (Darmstadt Institute for Hydraulic and Water Resources Engineering, Abrasion Test According to European Standard, EN 295 Part 3, 2016).

Liner Damage Although liner damage does not typically occur on the interior pipe surface, external impact damage may be displayed in the form of a star crack on the interior of the pipe. When observed, this condition should be repaired by qualified personnel in a manner recommended by the manufacturer, depending upon the severity and applicability to the manufacturing process and liner resin.

Leakage Depending upon the type of FRP encountered and the manufacturing process employed, leakage may exhibit itself in different ways. For pipe manufactured circa 1970, the leakage/failure may be catastrophic in nature. The complex phenomena causing this failure mode cannot be adequately explained within the parameters of this manual. Suffice it to say, industry concerns with this catastrophic failure mode have been addressed and rectified with current standards, testing, improved component material compatibility, and strength, as well as advancements in manufacturing methods. Leakage in fiberglass pipe manufactured post-1970s will most likely be observed in the joint by means of damaged or rolled rubber gaskets. Another cause of leakage may potentially be caused by gap and differing ovality between pipe barrels and pipe couplings, or integral bells caused by nonuniform bedding and backfill.

7.2.2 External Surfaces

Soil Corrosion Modern FRP is highly resistant to soils normally encountered in pipeline applications. However, care must be taken when encountering soils laden with high concentrations of hydrocarbons and other aggressive chemicals. If an adequate soil survey is performed prior to design and installation, consultation with pipe manufacturers and their resin suppliers can mitigate or eliminate the potential problem with the proper resin type selection. However, if discovered after installation, there are few, if any, measures that can be taken to repair or rehabilitate the host pipe.

Soil Stress Loading It is recognized that all soils creep over time. However, this condition seldom affects FRP.

Third-Party Damage Third-party damage may be encountered post-installation. As with any flexible buried pipe product, excavation related to new construction in the areas near an existing pipeline may cause direct or indirect damage to the pipe. The direct damage can potentially be impact damage. Indirect damage can occur in the form of excessive bending stress due to excavation or displacement of the soil support structure of the pipeline. Depending upon the severity of the incident, FRP pipelines may leak or fail. The prudent approach to third-party damage, when identified, is repair or replacement of the existing pipe.

Installation Damage Though not always accomplished, every pipeline installation should be professionally inspected. Typically, impact damage is caused during the handling and installation of the pipe. Installation damage tends to be located at the spigot or bell end of a fiberglass pipe segment, generally in the form of impact damage and/or cuts on the exterior surface of the pipe. These types of damage can be repaired per the manufacturer's recommended method. If the identified area is more severe and the structural integrity of the pipe is in question, then the damaged area of the pipe should be removed by cutting the pipe section. The manufacturer should be consulted for this type of repair.

As with any flexible pipe product, potential long-term damage can occur if inadequate or improper bedding and backfill is placed within the pipe zone. This type of danger can be avoided by the proper trench condition and pipe zone embedment placement in accordance with the available installation practices for fiberglass pipes such as the ASTM D3839 and the AWWA M45 "Fiberglass Pipe Design."

7.2.3 Complete Pipe Wall

FRP pipe structural walls can experience dry spots and delamination caused by a resin-starved environment occurring as a result of a manufacturing defect. As the FRP manufacturing methods have become more automated and accurate over the past 40 years of FRP development, this type of defect would typically only be seen in pipe produced prior to the 1980s. Such a condition is easily identified and seldom advances past standard in-plant quality control inspection. If observed, the resin-starved location should be removed, repaired, or the pipe section eliminated, depending on severity.

7.3 INSPECTION PROCEDURES

As with any pipeline material, there are two readily recognized conditions that warrant further inspection of an FRP pipeline: leakage and general water loss. When water loss of any sizable quantity is recognized, it is usually of a volume that is visible on the surface of the pipeline route. Also available for all leak detection on all pipe materials are newer leak detection technologies such as ultrasonic leak detection.

Man Entry In large diameter water pipes, when the pipeline can be completely shut down and taken out of service for an extended period of time, man entry into the pipeline is one possible inspection option. OSHA safety practices must be employed at all times. Valuable information may be obtained regarding the condition of the external surface and composite wall structure of the pipe by visual inspection of the interior surface.

For assistance in proper identification of damage to FRP pipe, ASTM D2563, the Standard Practice for Classifying Visual Defects in Glass-Reinforced Plastic Laminate Parts, is available for assistance. This standard covers acceptance criteria for visual inspection of parts made from glass-reinforced plastic laminates. It presents various descriptions of possible defects to serve as a guide for inspection of FRP products. This standard also categorizes different inspection requirements for levels of product quality. The acceptance criteria set in this standard can be applied to various material pipe properties such as dimensions and tolerances, inserts, molded-in threads or cored holes, workmanship, allowable visual defects, repairable defects, surface finish, and surface appearance.

Laser, Sonar, and Closed-Circuit Technologies These technologies can aid in the investigation of the FRP pipe. Mapping and other inspection practices are also suitable for use with FRP. Many of these technologies are rapidly emerging as revolutionary tools to the professional in nondestructive in situ pipeline inspection and evaluation.

Acoustical Emissions Inspection The acoustical emissions (AE) examination method can be used to detect damage to FRP pipe. The damage mechanisms detected in FRP are as follows: resin cracking, fiber debonding, fiber pullout, fiber breakage, delimitation, and bond or thread failure in assembled joints. Flaws in unstressed areas and flaws that are structurally insignificant will not generate AE.

This practice is convenient for online use under operating conditions to determine structural integrity of in-service FRP usually with minimal process disruption. Two ASTM standards for AE inspection are

- ASTM E1067, Standard Practice for Acoustic Emission Examination of Fiberglass Reinforced Plastic Resin (FRP) Tanks/Vessels, and
- ASTM E1118, Standard Practice for Acoustic Emission Examination of Reinforced Thermosetting Resin Pipe (RTRP).

Flaws located with AE should be examined by other techniques (for example: visual, ultrasound, and dye penetrate) and may be repaired and retested as appropriate. Repair procedure recommendations are outside the scope of this standard.

Differential Scanning Calorimetry Testing Differential scanning calorimetry (DSC) is a thermo-analytical technique in which the difference in the amount of heat required to increase the temperature of a sample and a reference are measured as a function of temperature. The main application of DSC is in studying phase transitions, such as melting, glass transitions, or exothermic decompositions. These transitions involve energy changes or heat capacity changes that can be detected by DSC with great sensitivity. DSC is used widely for examining polymers to check their composition.

Barcol Hardness Test This test method is most often used for composite materials such as reinforced thermosetting resins to determine the level of cure of a resin. The test complements the measurement of glass transition temperature as an indirect measure of the degree of cure of a composite. It is inexpensive and quick and provides information on the cure throughout a representative sample. The ASTM test method that can be referenced for this test is ASTM D2583. Minimum barcol hardness readings are specific to a given resin type, but they are usually approximately 30 at 75°F. Some FRP pipe manufacturers use additives in the liner resins to make the liner more flexible. This should be taken into consideration when determining the minimum barcol hardness reading to qualify a cured FRP product.

7.4 CONDITION DETERMINATION

To provide input to a rating system on the condition of an FRP pipeline, information obtained from the inspection should be analyzed both objectively and subjectively. Potential considerations to achieve this goal are as follows.

7.4.1 Schedule for Future Observation

If minimal interior damage (such as abrasion) is observed during an inspection, but it is determined to be of minimal structural concern or pipeline safety, then the pipeline should be scheduled for a future inspection. Regardless, the amount of abrasion should be quantified and compared with the original FRP pipe liner thickness to ensure that a sufficient liner thickness remains.

Note: The following "Rehabilitation" and "Replacement" sections are provided for information purposes only. They are not a component of the condition assessment process but are integral to an asset management program, and they were, therefore, included for reference.

7.4.2 Rehabilitation

Although FRP is a composite product, the bond between composite layers is formidable. As there are no metallic components, adding an exterior protection system (such as cathodic protection) to the pipeline is obsolete and of no value whatsoever. The same is true of the interior. Relining and/or slip lining with another thermoplastic material is of little to no value when compared with the investment.

7.4.3 Replacement

As with any pipeline, total and complete line replacement will be the choice of last resort. Such a radical conclusion should only be chosen when spot repairs of selected pipeline sections (such as those described next) have proven unsuccessful.

7.5 REPAIR PRACTICES

Different repair methods are available to the water works professional and contractor. The responsible party should select the option that best suits the problem and site condition.

7.5.1 Straight Pipe Section: Permanent Repair with External Coupler

FRP Coupler The versatility of some manufacturers' standard joining couplers allows these couplers to slide on the exterior of a pipe's surface. In these cases, a standard FRP coupler may be used to facilitate the joining of the replacement pipe section to the pipeline. This can be accomplished in the following manner:

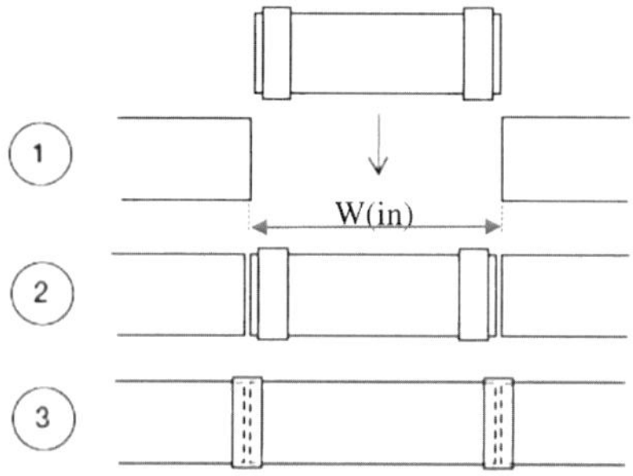

Fig. 7-1. Diagram of FRP coupler
Source: Courtesy of Flowtite Technology

1. Following observation and analysis of the damaged or leaking pipe section, decide upon the length of pipe to be removed (dimension W, as shown in Fig. 7-1).
2. Cut and remove this section of the pipeline and replace with a new section of pipe.
3. Inspect and examine the ends of the still installed pipe and the replacement pipe section to ensure a solid composite structure exists.
4. The exposed cut ends of both the existing pipe and the replacement section (dimension W) should be coated with resin to prevent the medium from attacking open glass fibers. The resin must be cured prior to placement in the pipeline.
5. Insert the replacement pipe (dimension W) into the pipeline.
6. Reconnect the system using two standard FRP couplers.
7. Consult with manufacturer to determine if the standard coupler is suitable for this procedure.

This procedure is applicable for unrestrained underground pipeline sections.

Mechanical Coupler

1. Following observation and analysis of the damaged or leaking pipe section, decide upon the length of pipe to be removed, dimension W (Fig. 7-2).
2. Cut and remove this section of the pipeline and replace with a new section of pipe.
3. Inspect and examine the ends of the still installed pipe and the replacement pipe section to ensure a solid composite structure exists.
4. The exposed cut ends of both the existing pipe and the replacement section (dimension W) should be coated with resin to prevent the medium from attacking open glass roving. The resin must be cured prior to placement in the pipeline.

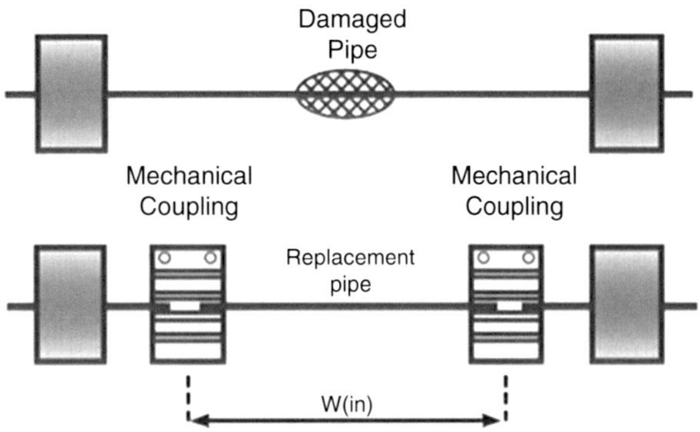

Fig. 7-2. Diagram of mechanical coupler
Source: Courtesy of Flowtite Technology

5. Insert the replacement pipe (dimension W) into the pipeline.
6. Reconnect the system using two mechanical couplers (e.g., Straub, Dresser, Victaulic Depend-O-Lok, Ford, Smith-Blair, or equivalent).

This procedure is applicable for unrestrained underground pipeline sections.

Butt and Wrap Field Lamination This procedure is applicable for either unrestrained or restrained aboveground or underground pipeline sections.

1. Following observation and analysis of the damaged or leaking pipe section, decide upon the length of pipe to be removed, dimension W (Fig. 7-3).

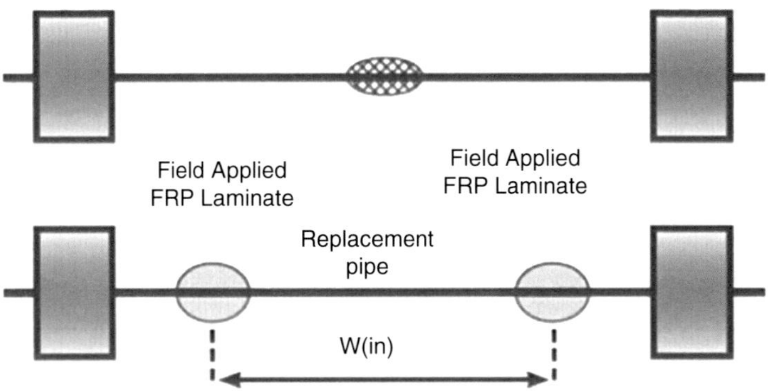

Fig. 7-3. Diagram of butt and wrap field lamination
Source: Courtesy of Flowtite Technology

2. Cut and remove this section of the pipeline and replace with a new section of pipe. Inspect and examine the ends of the still installed pipe and the replacement pipe section to ensure a solid composite structure exists.
3. The exposed cut ends of both the existing pipe and the replacement section (dimension W) should be coated with resin to prevent the medium from attacking open glass roving. The resin must be cured prior to placement in the pipeline.
4. Insert the replacement pipe (dimension W) into the pipeline. The ends of the pipes must be prepared appropriately, in accordance with the pipe manufacturer's recommendation, with light sanding and cleaning prior to the application of the field butt and wrap.
5. Reconnect the system using two butt and wrap field lamination kits provided by the manufacturer of the replacement pipe.

7.5.2 Emergency Repair by Laminating

If circumstances prohibit installation of a complete and permanent repair, a temporary repair may be used.

One temporary repair method is covering the defect with a field laminate kit (similar to the butt and wrap field lamination kit). The pipeline internal pressure must be eliminated. The area must be dry. The area around the damage should be removed with a grinder. The ground surface should be thoroughly dry. The resin and hardener shall be applied with a brush. Then, alternating layers of glass fiber and resin should be applied, rolled to remove air, and allowed to thoroughly cure before reinstating the pipeline pressure. The manufacturer should be consulted for recommendation as to the length and thickness required for proper repair of each situation.

7.5.3 Emergency Repair by Cementing a Pipe Segment

The pipeline internal pressure must be eliminated. The area must be dry. The area around the damage should be removed with a grinder. The ground surface should be thoroughly dry. The resin and hardener should be applied to the interior and exterior of the damaged area via roller or brush. Alternatively, Sikabond may be used.

1. Position an FRP saddle over the area to be repaired.
2. Fix the saddle to the pipe with tensioning belts until completely hardened.
3. Reinstate pressure. The manufacturer should be consulted for recommendation as to the length and thickness required for proper repair of each situation.

7.5.4 Emergency Repair by External or Internal Clamp

A simple method to stop leakage is by installing a rubber-lined clamp to either the exterior or interior surface. There are several commercially available sources of the product (e.g., Dresser, Depend-O-Lok, Weko-Seal, Ford Meter Box, and Smith Blair). It is recommended that once the leak has been stopped by clamping, a fiberglass (or other suitable material) lay up should be applied over the clamp to prevent corrosion.

REFERENCES

ASTM. (2011). "Standard practice for acoustic emission examination of fiberglass reinforced plastic resin (FRP) tanks/vessels." *ASTM E1067*, West Conshohocken, PA.

ASTM. (2013). "Standard test method for indentation hardness of rigid plastics by means of a barcol impressor." *ASTM D2583-13a*, West Conshohocken, PA.

ASTM. (2014). "Standard guide for underground installation of 'fiberglass' (glass-fiber reinforced thermosetting-resin) pipe." *ASTM D3839*, West Conshohocken, PA.

ASTM. (2015). "Standard practice for classifying visual defects in glass-reinforced plastic laminate parts." *ASTM D2563-08*, West Conshohocken, PA.

ASTM. (2016). "Standard practice for acoustic emission examination of reinforced thermosetting resin pipe (RTRP)." *ASTM E1118*, West Conshohocken, PA.

AWWA (American Water Works Association). (2013). "Fiberglass pipe design." *AWWA M45*, Denver.

Darmstadt Institute for Hydraulic and Water Resources Engineering. (2016). "Abrasion test according to European Standard." *EN 295 Part 3*, Darmstadt, Germany.

CHAPTER 8
PIPE MATERIALS: POLYVINYL CHLORIDE

8.1 INTRODUCTION

It is difficult to inspect installed polyvinyl chloride (PVC) pressure pipe because of limited access. Inspection is typically limited to indirect methods such as

- Review of water loss records,
- Review of breakage data, and
- Pressure testing of sections of pipeline.

Excessive water loss and/or breakage records could indicate that a problem exists and that a condition assessment program would be appropriate.

If the review identifies a problem, possible assessment techniques include

- Closed-circuit television (CCTV) inspection of the inside of the pipe,
- Excavation for visual inspection of the outside of the pipe, and
- Forensic testing.

8.1.1 PVC Pipe Properties

PVC pipe is a nonconductor of electricity and is immune to electrochemical reactions caused by acids, bases, and salts. PVC is well-suited for all normal water chemistries across the United States and Canada and is resistant to galvanic corrosion for both the inside and outside of the pipe.

PVC pipe is made from rigid (unplasticized) PVC compound. PVC pipe is resistant to fracture because of its ability to yield under loading without breaking. For external dead and live loads, PVC pipe functions as a flexible pipe. For internal pressure loads, PVC's flexibility results in relatively low wave velocity and consequently lower pressure surges (i.e., water hammer).

As with all pipe materials, any excavation for pipeline access, sampling, or verification must be performed with care to prevent damage.

8.1.2 Joints

Joining systems commonly used in the installation of PVC pipe for municipal applications are integral-bell gasketed joints, pin-and-groove gasketed joints, spline-lock gasketed joints, internally restrained gasketed joints, and butt-fused joints.

Nonrestrained Joints

Integral-bell gasketed joint. By far the most common joint type for municipal pipe, this joint is characterized by a small exterior hump in the bell for the gasket raceway (Fig. 8-1). Although this joint type is nonrestrained, restraint can be achieved by installation of external restraint devices after pipe joint assembly.

Restrained Joints

Pin-and-groove gasketed joint. Adjacent to the bell's gasket raceway, there is a ring with a series of holes. Each hole contains a pin that passes through the ring into a hole in a similar ring in the pipe spigot (Fig. 8-2).

Spline-lock gasketed joint. This joint uses a gasketed coupling with nylon splines extending from holes in the coupling (Fig. 8-3). The splines extend into aligned grooves in the coupling and the pipe spigot.

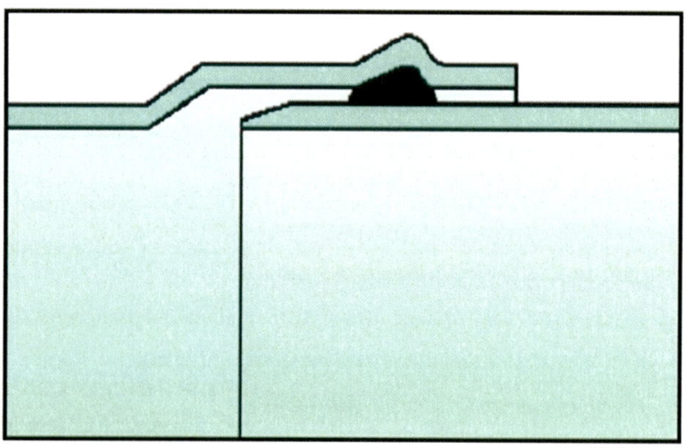

Fig. 8-1. Integral-bell gasketed joint
Source: Courtesy of Uni-Bell PVC Pipe Association

Fig. 8-2. Pin-and-groove gasketed joint
Source: Courtesy of Uni-Bell PVC Pipe Association

Fig. 8-3. Spline-lock gasketed joint
Source: Courtesy of Uni-Bell PVC Pipe Association

Internally restrained gasketed joint. This joint's bell has a much more extensive gasket raceway, because there needs to be room to accommodate the gasket and an adjacent metal restraining structure (Fig. 8-4). The wide raceway is diagnostic of this joint type.

Butt-fused joint. This type has no bell, but instead an interior and exterior bead at each joint (Fig. 8-5).

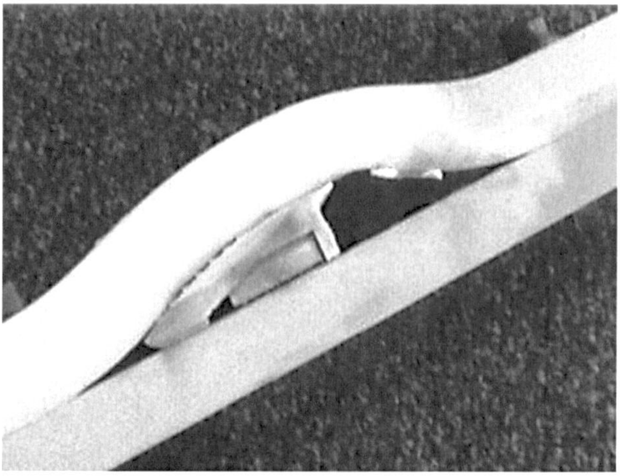

Fig. 8-4. Internally restrained gasketed joint
Source: Courtesy of Uni-Bell PVC Pipe Association

Fig. 8-5. Butt-fused joint
Source: Courtesy of Uni-Bell PVC Pipe Association

8.2 HISTORY

Polyvinyl chloride was discovered in the last decade of the nineteenth century. However, the material at that time was so resistant to modification that it could not be easily processed into useful applications.

It was not until the 1920s that scientists in Europe and America launched an extended campaign to commercialize PVC plastics. PVC technology continued to evolve worldwide, but particularly in Germany. In 1931, extrudable PVC compounds were developed by German scientists, and millions of pounds of PVC were produced, some for pipe. In large part, the PVC pipe industry was born out of a necessity created by World War II, when metallic materials were needed for the war effort. German scientists and engineers turned to PVC, using it as a pipe material to quickly restore essential water and wastewater pipelines in cities damaged by Allied bombings. Several of these earliest PVC pipelines are still in service.

In the two decades following World War II, there were considerable advances in PVC technology, both in formulation and processing. During this time, the development of more effective stabilizers, lubricants, and processing aids, together with processing machinery engineered specifically for PVC, allowed for increasing success in the extrusion of PVC pipe.

PVC pipe was introduced to North America in 1951. In 1955, ASTM organized a group to write plastic pipe standards. The American Water Works Association (AWWA) followed in the 1960s. Current ASTM and AWWA standards for PVC pressure pipe and fittings are

- ASTM D1785—"Standard for Polyvinyl Chloride (PVC) Plastic Pipe, Schedules 40, 80, and 120"—first published in 1960 (currently to 24 in.)
- ASTM D2241—"Standard for Polyvinyl Chloride (PVC) Plastic Pressure-Rated Pipe (SDR PR Series)"—first published in 1964 (currently to 36 in.)
- ASTM F1483—"Standard for Oriented Polyvinyl Chloride (PVCO) Pressure Pipe"—first published in 1993 (currently to 16 in.)
- AWWA C900—"Standard for Polyvinyl Chloride (PVC) Pressure Pipe and Fabricated Fittings, 4 in. through 60 in."—first published in 1975
- AWWA C907—"Standard for Injection-Molded Polyvinyl Chloride (PVC) Pressure Fittings, 4 in. through 12 in."—first published in 1991 (to size 8 in.)
- AWWA C909—"Molecularly Oriented Polyvinyl Chloride (PVCO) Pressure Pipe, 4 in. through 24 in."—first published in 1998 (to size 12 in.)

8.3 CONDITION ISSUES

8.3.1 Internal Surfaces

Internal Corrosion Because PVC pipe does not electrochemically corrode, there is no tuberculation caused by corrosion by-products.

Chemical Exposure In general, the interior surfaces of PVC piping systems are not adversely affected by chemicals found in typical potable water and reclaimed water systems. There is no record of oxidation-induced failure in PVC water distribution pipes. The resistance of PVC pipes to water disinfectants has been determined through accelerated aging studies per ISO 4433 (8 ppm ClO_2 at 40°C).

Potability PVC water pipe is listed to NSF Standard 61, "Drinking Water System Components—Health Effects." However, for PVC pipe produced prior to 1977, there have been cases of very small diameter pipe exceeding the revised EPA vinyl chloride monomer limits. For this situation to occur, the pipelines must have long retention times and have warm water temperatures. A simple water chemistry test should determine whether this issue exists.

8.3.2 External Surfaces

External Corrosion Because PVC is a nonconductor, galvanic and electrochemical attack does not occur in PVC pipe and fittings. PVC suffers no damage from attack of normal or corrosive soils.

Chemical Attack Resistance of PVC pipe and elastomeric gaskets to attack by chemical agents has been determined through years of research and field experience. Pipe and gaskets are able to withstand a broad range of acidic and caustic environments. Typically, either prolonged or repeated exposure is necessary to compromise the pipe system. One source of information on the chemical resistance of PVC pipe is the National Association of Corrosion Engineers publication entitled, "Corrosion Data Survey, Nonmetals Section."

Soil Movement PVC pipe has considerable bending flexibility as well as a bell and spigot joint that allows some stress-free longitudinal movement. This combination allows effective performance while permitting some ground movement. Failures such as breaks from soil movement, minor earthquakes, and expansive clay soils are not frequent occurrences for PVC pipes. Excessive movements that do cause breakage or joint separation will result in leaks that are obvious for condition assessment.

Ultraviolet Radiation Exposure to ultraviolet (UV) radiation results in a change in a PVC pipe's surface color and some reduction in impact strength. Other properties such as tensile strength (pressure rating) and modulus of elasticity (pipe stiffness) are not seriously affected. There are no industry restrictions on cutting or tapping of UV-discolored PVC pipe.

8.3.3 Material Considerations

Temperature Limitations PVC is recommended for applications in which temperatures of the pipe wall do not exceed 140°F. Per Uni-Bell's *Handbook of PVC Pipe*, derating of the pipe's strength is recommended from 80°F to 140°F.

Cyclic Surge Pressures All pipe materials can experience fatigue failure when exposed to cyclic surge pressures that exceed design limits. However, fatigue failure due to cyclic surging is typically not a matter of concern for PVC water transmission and distribution systems; these systems rarely experience cyclic surging with frequency and magnitude high enough to influence pipe performance. Naturally, when surges occur, they should be addressed if they have the potential to exceed the pipe's design limits. When excessive cyclic surging has occurred, the warning signs are very short slits or cracks on the exterior pipe surface. If leakage through these slits is not observable, microscopic evaluation may be required.

Air exists in all water pipelines. According to AWWA Manual M51, there are many sources of air. One is the water itself, which contains at least 2% dissolved air by volume in standard pipeline operating conditions. Air collects in pockets at high points along the pipeline and can be detrimental if not controlled. Air can magnify surge pressures, which can damage the pipe and pipeline components and can create unsafe conditions for tapping of pipe. Proper air release is important for prevention of cyclic surge problems for all pipe materials, including PVC. Condition assessment of a pipeline should include investigation of air valves.

Modulus of Elasticity For PVC pressure pipe, the modulus of elasticity is a minimum of 400,000 psi. The modulus is used in calculations for pipe deflection and for surge pressures. This material property remains constant over time, so it can be used for analysis by an assessor.

Crack Propagation For PVC pipe, in most cases, the fracture or crack is relatively short and, at the maximum, will be limited or arrested by the gasketed bell and spigot joint.

Rapid crack propagation (RCP) is a fast, long running fracture. RCP is a postfailure event, requiring an initiation of a fracture and an energy source to drive the fracture in the pipe wall in the direction of the pipe axis. In a properly installed and maintained pipeline, the pressures required to extend a crack are well above the pipe pressure rating. RCP may run for multiple pipe lengths when joints are butt fused or solvent cemented together. RCP generally occurs because of a combination of both improper installation practices as well as increased pressure and dynamic loading due to excessive air in the line. Proper pipe design and installation methods in combination with proper air removal practices should always be followed to prevent RCP.

8.3.4 General

Permeation According to an AWWA Research Foundation report, titled "Impact of Hydrocarbons on PE/PVC Pipes and Pipe Gaskets," PVC pipe with SBR gaskets may be used (without exceeding the MCL for benzene) for water mains in all conditions of gasoline contamination provided there is at least a minimal flow in the main. It should be recognized that

1. Some stagnant or no-flow conditions may exist, and
2. The source research states that it did not test all manner of exposure and strained conditions for adverse and potential effects on physical properties.

PVC pipe itself is impervious to gasoline—any gasoline permeation that takes place is through the gaskets.

Abrasion Research studies in Europe and North America on PVC pipe's abrasion resistance have shown that PVC pipe will erode over a broad area rather than develop localized pitting.

Third-Party Damage Like all materials, PVC is subject to construction-related damage. This type of problem is usually recognizable by ground seepage or high levels of water loss.

Overbelling of Gasketed Joints An insertion line is located on the exterior of each pipe spigot. A pipe spigot that has been inserted into a bell past the insertion line may contact the inside of the bell, causing stress that has not been included in the joint design. The stress in the bell is not of great concern because of PVC pipe's tendency for stress relaxation over time. If an assessor is not able to view the line at the face of the bell, the bell may be overinserted. This is generally not a cause for leakage as the stress relaxation function usually keeps the gasket under adequate compression.

Overdeflection of Gasketed Joints (Longitudinal) Each pipe manufacturer has published allowable longitudinal deflection for its pipe joints. A pipe spigot that has been deflected more than the allowable parameters may contact the inside of the bell, causing stress that has not been included in the joint design. Again, PVC pipe's tendency for stress relaxation will generally allow any stress that has developed to dissipate over time.

It is difficult for an assessor to determine whether a joint has been overdeflected because the deflections are very small (on the order of 1 to 3°).

Again, this is generally not a cause for leakage as the stress relaxation function usually keeps the gasket under adequate compression.

8.4 INSPECTION PROCEDURES

PVC pipe can be inspected for its condition assessment using several methods such as water loss surveys, internal inspection using CCTV and cameras, man entry for larger pipe inspection, and leakage detection using acoustical monitoring. These techniques may be helpful in identifying leaks at joints, connections, and fittings.

8.5 CONDITION DETERMINATION

A condition determination for PVC pipe can be done by examining the breakage and leakage rates as a predictor of future performance. PVC is not subject to electrochemical corrosion, so there is not a mechanism for material degradation over time. If the PVC pipes' breakage and leakage rates are greater than acceptable limits, then rehabilitation or replacement may be needed.

Note: The following rehabilitation section is provided for information purposes only. It is not a component of the condition assessment process but is integral to an asset management program, and it was, therefore, included for reference.

8.6 REHABILITATION PRACTICES

Rehabilitation of PVC can be done by several methods including slip lining, internal joint seals, and/or the use of external clamps. The nominal outside diameter will be marked on the print line of the pipe. PVC pipes can have either IPS or cast iron outside diameters. Repair couplers and clamps are readily available for both IPS and cast iron diameters from local waterworks distributors.

REFERENCES

ASTM. (1960). "Standard for polyvinyl chloride (PVC) plastic pipe, schedules 40, 80, and 120." *ASTM D1785*, West Conshohocken, PA.

ASTM. (1964). "Standard for polyvinyl chloride (PVC) plastic pressure-rated pipe (SDR PR series)." *ASTM D2241*, West Conshohocken, PA.

ASTM. (1993). "Standard for oriented polyvinyl chloride (PVCO) pressure pipe." *ASTM F1483*, West Conshohocken, PA.

AWWA (American Water Works Association). (1975). "Standard for polyvinyl chloride (PVC) pressure pipe and fabricated fittings, 4 in. through 60 in." *AWWA C900*, Denver.

AWWA. (2016). "Air valves: Air release, air vacuum, and combination." *AWWA M51*, Denver.

AWWA. "Molecularly oriented polyvinyl chloride (PVCO) pressure pipe, 4 in. through 24 in." *AWWA C909*, Denver.

AWWA. "Standard for injection-molded polyvinyl chloride (PVC) pressure fittings, 4 in. through 12 in." *AWWA C907*, Denver.

AWWA Research Foundation. (2008). *Impact of hydrocarbons on PE/PVC pipes and pipe gaskets*, Alexandria, VA.

ISO. (1997). "Thermoplastics pipes—Resistance to liquid chemicals—Classification." *ISO 4433*, New York.

National Association of Corrosion Engineers. (1985). *Corrosion data survey, nonmetals section*, Houston.

NSF/ANSI (National Science Foundation/American National Standards Institute). "Drinking water system components—Health effects." *NSF/ANSI 61*, Arlington, VA.

Uni-Bell. (2012). *Handbook of PVC pipe*, Industrial Press, New York.

CHAPTER 9
PIPE MATERIALS: ASBESTOS CEMENT

Asbestos cement (AC) pipe is made from a mixture of portland cement (or cementitious materials) and asbestos fibers, with or without silica. AC pipe installed in North America was typically manufactured to ASTM C500 specifications as either Type I or Type II pipe. The first AC pipe was introduced and used in the United States circa 1930 and was Type I pipe, which consisted of approximately 80% portland cement and 20% asbestos fibers. In Type II AC pipe, introduced in the United States in the mid-1930s, 40% of the portland cement was replaced with silica and the pipe was autoclaved. The addition of silica, combined with the autoclave curing process, caused the free lime to form more stable compounds within the pipe. As a result, the Type II pipe proved more resistant to acids and sulfates; it is also considered chemically resistant pipe, when tested in accordance with methods of ASTM C500. Type II pipe became the predominant type of pipe used in North America from the 1940s forward.

AC pipe had been mainly used for potable water mains of 24 in. and less in diameter as opposed to larger water mains and transmission lines. It can be drilled and tapped for connecting but does not have the same strength or suitability for threading as steel, cast iron, or ductile iron, and any leakage in the thread will become worse as time passes. Other applications of AC pipe have been for gravity sewer and storm drainage, force mains, and industrial effluent and process piping.

Failure modes of AC pipe may vary but generally include circumferential breaks, longitudinal splits, joint failure, and softening of the pipe wall that can lead to blowouts. Type II, chemical resistant AC pipe, is more likely to experience circumferential breaks, longitudinal splits, or joint failures. The most common of these failure modes is circumferential breaks. Nonetheless, any one of these failure modes would likely lead to a catastrophic failure.

AC pipe is relatively brittle and particularly susceptible to failure in locations subject to significant ground movement.

Because of its smaller diameters, AC pipe inherently lends itself to fewer inspection techniques, whereas the larger pipelines are often conducive to manned entry, allowing internal visual inspection techniques. The most reliable determination of AC pipe deterioration and mechanical properties requires excavation of the pipe. Other testing methods such as acoustical testing can also be used. The appropriate condition assessment approach for AC pipe is largely dependent on the failure history and failure modes experienced by the owner of the system.

9.1 HISTORY

Asbestos cement pipe was introduced into the North American water industry in 1931. It had been used just a few years in some European countries prior to its North American introduction. AC pipe is a mixture of portland cement, or cementitious materials, and asbestos fibers (with or without silica). The installation of AC pipe was peaking between the 1950s through the 1980s, when health concerns (related to asbestos fiber exposure during the manufacturing process) forced the discontinued use of AC pipe in the United States.

In some U.S. cities, AC pipe still accounts for a large portion of the water distribution pipe material. An estimated 1.4 million mi of AC pipe was in service by 1974, 200,000 mi of which were in the United States alone. AC pipe built and installed in the 1930s still exists in operation today in pressurized potable water systems.

AC pipe performance has shown that it typically fails as a result of stresses and strains in the pipe wall exceeding the strength of the pipe, which is reduced as the pipe deteriorates by internal and/or external chemical attack. Another external influence compromising pipe integrity is unstable soil conditions related to settling, expansive soils, or poor backfill and compaction.

Failure modes of AC pipe include longitudinal cracking, circumferential breaks, joint gasket failures, deterioration, and loss of wall section (resulting in holes in the pipe wall and a *softening effect* of the pipe). Deterioration is typically caused by external and/or internal corrosion from acidic soils or water, for example, high soluble sulfates in soil or water.

AC pipe wall thickness generally increases as the pipe diameter increases. Consequently, smaller diameter pipe is more susceptible to deterioration and has lower mechanical strength than larger diameter pipe. Small diameter AC pipe will usually exhibit higher failure rates than larger diameter pipe of similar initial chemical composition, age, and environmental exposure.

Smaller diameter pipes are more prone to failures resulting from longitudinal stresses, due to differential settlement, than from internal water pressure. Studies in England and Canada indicate AC pipe failure rates of approximately six to 10 breaks per 100 km of pipeline per year of service (or nine to 16 breaks per 100 mi of pipeline per year). These failure rates increased with the age of the pipe.

9.1.1 Manufacture

Type I AC pipe in the United States was manufactured using a mixed aqueous slurry of portland cement (80–85%) and a mixture of relatively long and medium grade chrysotile asbestos fibers (15–20%). Type II AC pipe was manufactured using a mixed aqueous slurry of portland cement (40–45%) and a mixture of relatively long and medium grade chrysotile asbestos fibers (15–20%) and silica (40%), which was added to the mixture to make the pipe corrosion resistant. The slurry was dewatered by a machine using a rotary sieve cylinder and a continuous felt to produce a very thin layer of AC, which was wrapped around a mandrel under pressure until a pipe with the desired wall thickness was produced. The mandrel was then extracted, and the pipe was cured (1) by passing through a tunnel-like low temperature oven followed by immersion in, or spraying with, water (Type I); or (2) by autoclaving (Type II). After curing, the ends of the pipe lengths were cut and finished to receive couplings that were produced by cutting larger diameter pipe into sections. AC pipe was manufactured under carefully controlled conditions to achieve a smooth internal finish. The encapsulated asbestos fibers acted as a reinforcing material to form an intrinsic part of the cement matrix.

9.1.2 Field Joints

When installing AC pipe, two types of joints were predominantly used: a cast iron detachable joint and an AC coupling joint. Leaded joints poured with a ladle, although not common, were also used in the 1930s to join AC pipe to cast iron fittings.

The cast iron detachable joint consists of two cast iron flanges, a cast iron central collar, and two rubber rings, along with a set of nuts and bolts for a particular joint. For this joint type, the AC pipe would have flush ends. For jointing a flange, a rubber ring and a collar were slipped to the first pipe, then a flange and a rubber ring were introduced from the jointing of the next pipe. Both pipes were then aligned, the collar centralized, and the joints of the flanges tightened with nuts and bolts.

The AC coupling joint consisted of an AC coupling and two or three rubber rings. The pipe ends for the AC coupling were machined and chamfered. The rubber rings were positioned in the grooves inside the coupling. Next, pipe lubricant was applied to the chamfered end of the pipe and the coupling, and the pipe was pushed with the help of a bar pressed against a block of wood (positioned at the pipe end) or by using a backhoe bucket against a wood block against the pipe end. Because the wall thickness of the machined ends was reduced, the machined ends could be more vulnerable to failure if they extended beyond the coupling.

9.2 CONDITION ISSUES

9.2.1 External Surfaces

Corrosion Soils conducive to lime leaching and corrosive to AC pipe typically have high concentrations of sulfates, low pH, and a high or variable groundwater table in the pipe zone. The primary mechanisms leading to chemical degradation from aggressive chemicals in the surrounding soil are sulfate deterioration and acid attack. High levels of soluble sodium in the area may indicate that the pipe bedding area contains significant amounts of sodium sulfate. Soluble sodium sulfate can induce sulfate deterioration in concrete, resulting in a softening of the pipe wall. Generally, concrete is not resistant to acid attack and can also degrade in the presence of bicarbonates at neutral pHs.

Soil moisture content may also influence the external corrosion rate. Wetter soils are typically more conducive to exterior deterioration of AC pipe than dry or fast draining soils.

When an area of the wall loses sufficient strength, the internal pressure can result in blowouts of the pipe wall. These blowouts are typically catastrophic in nature and occur without warning (Fig. 9-1).

Joint Failure Joint gaskets have been known to dry out and lose elasticity, resulting in a release of water. Within hours, the pipe wall is eroded from inside out, often resulting in complete separation of the joint and catastrophic failure. Another type of joint failure is due to excess deflection at the joint, resulting in a leak that eventually erodes the pipe wall from inside out and blowing out the joint (Fig. 9-2).

Circumferential Breaks Circumferential breaks, often referred to as beam breaks and back breaks, are typically caused by longitudinal stresses and are one of the most common failure modes of AC pipe. Longitudinal stresses are associated with connections to structures, unstable soils, clayey and/or expansive soils, a void in the soil, poor bedding, poor installation, and third-party influences (e.g., adjacent utilities and/or incidental work in the pipe bedding area). In addition, corrosion attack on the pipe can reduce structural integrity, making the pipe more susceptible to these stresses and resulting in failure (Fig. 9-3).

Fig. 9-1. Catastrophic AC main blowout due to corrosion
Source: Courtesy of Tucson Water

Fig. 9-2. Coupling gasket joint failure
Source: Courtesy of Tucson Water

Longitudinal Breaks Longitudinal breaks of AC pipe are not as common as circumferential breaks. A failure mode of this type is often associated with hoop stresses related to pressure in the pipe, soil overloads, live loads, and frozen moisture in the ground (Fig. 9-4).

Fig. 9-3. Circumferential break
Source: Courtesy of Tucson Water

Fig. 9-4. AC pipe longitudinal break/failure
Source: Courtesy of Tucson Water

Third-Party Damage Excavation related to new construction may damage an existing pipeline. Generally, in the case of AC water lines, such damage takes the form of nicks, scrapes, or settling and often does not necessarily involve an immediate leak or failure. Direct impact by horizontal drilling installations, however, may necessitate an immediate repair (Fig. 9-5).

Fig. 9-5. Third-party damage, borehole at invert
Source: Courtesy of Tucson Water

Installation Errors During installation of AC pipe, gaskets have been pinched or rolled (often referred to as *fish mouth*) and the problem has not been discovered until several years, or even decades, later. This type of failure is considered joint failure, albeit the root cause of failure is installation error. Other installation errors are poor or inadequate bedding. These types of construction practices and installation errors often contribute to circumferential breaks and joint failures.

Internal Corrosion AC pipe can also corrode from the inside. Soft or acidic water can be particularly damaging to AC pipe. The Langelier index (LI) and aggressive index (AI) can be used to assess the aggressiveness of potable water transported through the pipe. ASTM C500/C500M provides useful guidelines and test methods for determining remaining service life of AC pipe influenced by various levels of aggressive soil environments. The AI taken from ASTM C500 is a measure of the corrosivity of water toward AC, and the following is an example of the usefulness of this standard:

$$\mathrm{LI} = \mathrm{pH} + \mathrm{pH_S}$$

$$\mathrm{AI} = \mathrm{pH} + \log(\mathrm{AH})$$

where

pH_S = pH at which water is saturated with calcium carbonate
A = alkalinity (milligrams per liter of calcium carbonate)
H = hardness (milligrams per liter as calcium carbonate)

The relative aggressiveness of water is defined in Table 9-1.

Failure Patterns Many water utilities and recent studies have reported that after an AC failure had occurred at a particular location, subsequent failures had occurred in the same vicinity. These subsequent failures may be indicative that the pipe in a particular area is nearing the end of its service life. However, the subsequent failures could also be due to the additional hydraulic stresses from the initial failure or from disturbances from the repair and cleanup efforts. Testing in accordance with the ASTM C500 methods may help conclude the root causes and remaining useful life.

9.3 INSPECTION PROCEDURES

Most condition assessment techniques of AC pipe require excavating, exposing, and removing samples of the pipe. Testing and assessment methods focus on testing and examining the internal and external pipeline environment, i.e., aggressiveness of the soil and water surrounding the pipe and the aggressiveness of the water conveyed in the pipe. The tools and techniques used to perform these tests include, but are not limited to, mechanical tests, chemical tests, hardness tests, visual tests, and acoustic tests.

Soil and Groundwater Testing around Pipe For AC pipe, the primary mechanisms for potential chemical degradation from aggressive chemicals in the surrounding soil are sulfate deterioration, concrete carbonation, and acid attack. Testing will reveal potential problems related to leachable calcium, sodium, and sulfate ions in the surrounding soil and groundwater.

Table 9-1. Relative Aggressiveness of Water

	Aggressiveness		
Index	High	Moderate	Low
Langelier	<-2	-2 to 0	>0
Aggressive	<10	10 to 11.9	≤ 12

Conveyed Water Testing of Pipe The aggressiveness of the conveyed water can result in degradation of AC pipe. For example, soft water with very low carbonate and bicarbonate content could result in the leaching of free lime from the cement. Testing of the conveyed water in the pipe can indicate how aggressive the water is measured by using the AI. Internal and external waters are known to cause aggressive environments, and soluble sulfate contents of water and soils are considered separately even though they may exist in combination (ASTM C500/C500M).

Mechanical Testing of Pipe Pipe samples from failed pipe and coupons from in-service pipe can be mechanically tested. Commonly used mechanical tests include the hydrostatic pressure test, to measure the burst strength; the flexural test, to determine whether the pipe can withstand the loads stated in the specifications; the crushing test, to determine whether the pipe can withstand the crushing loads stated in the specifications; and Schmidt hammer tests, to measure the elastic properties or strength of the AC pipe (mainly surface hardness and penetration resistance). O-ring condition can be tested by applying a compression test (ASTM D395), a hardness test (ASTM D1415), and Fourier transform infrared (FTIR) spectroscopy.

Chemical Testing of Pipe and Gaskets Chemical tests can be performed on both the AC joint gasket and the pipe to identify degradation of AC pipe and gaskets. For example, the FTIR spectroscopy can be used to identify the purity and structure of rubber gaskets, indicating chemical loss or breakdown. Phenolphthalein can be used to identify the degradation of AC pipe due to leaching of calcium.

Electron Microscopes Scanning electron microscope (SEM) and energy dispersive spectroscopy (EDS) are two examples of microscopes that use a focused beam of electrons to analyze chemicals, e.g., AC pipe samples to determine pipe degradation (Fig. 9-6).

Phenolphthalein Dye Testing Phenolphthalein dye testing can be used to identify how much of the pipe cross section has degraded (carbonated). When phenolphthalein dye is injected or sprayed over the cross section of an AC coupon, the cross section will turn purple or pink where calcium is still available, whereas the white area may indicate the areas where calcium has leached from the pipe (Fig. 9-7).

Chemical Analysis Using EDS EDS with an electron microscope can be used to determine the chemical composition of AC pipe segments. Fig. 9-8 depicts the elemental distribution of multiple pipe segments. Careful analysis using EDS can help identify relationships between failed

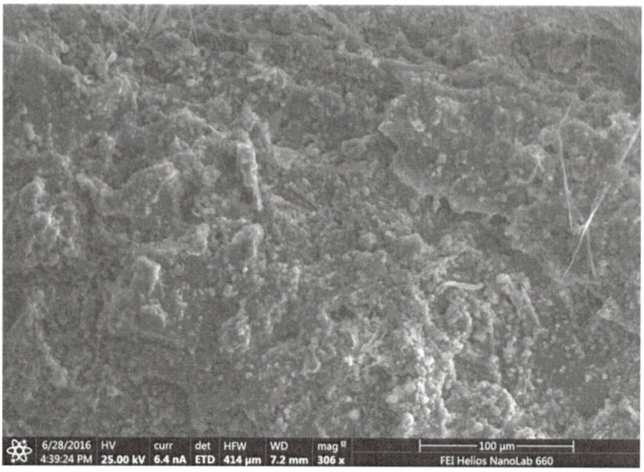

Fig. 9-6. Scanning electron microscope (SEM) of AC pipe sample
Source: Courtesy of Tucson Water

Fig. 9-7. Phenolphthalein dye test of asbestos cement pipe wall cross section
Source: Courtesy of Tucson Water

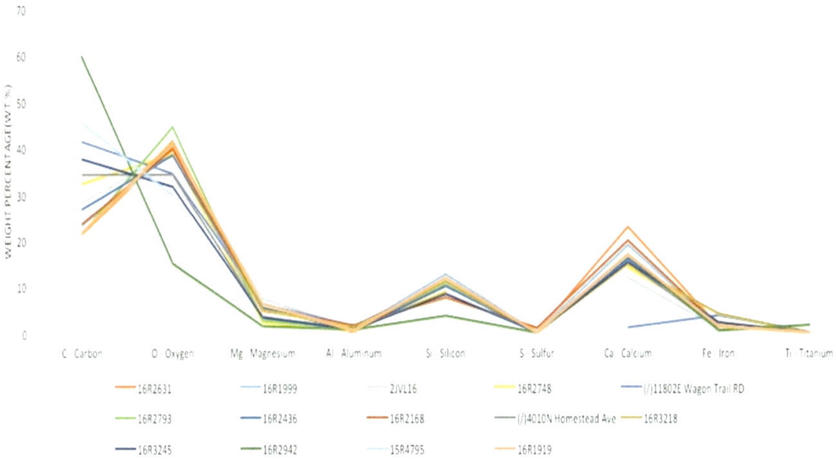

Fig. 9-8. Energy dispersive spectroscopy of multiple pipe samples
Source: Courtesy of Tucson Water

pipe and the failed pipe's elemental composition. This information can be helpful in predicting remaining service life of pipe cohorts of similar composition.

Sonic Inspection Sonic inspection entails sending a receiver through a pressurized pipeline and listening for acoustic events that may indicate leaks in the pipe wall or pipe joint. The level of the sound or frequency can help evaluate the size of the leak. The approximate location of the leak can also be determined by correlation methods and correlating equipment. However, leaks in AC pipe typically manifest themselves quickly as pressurized fluid erodes the pipe or coupling wall away.

Acoustic Pipe Wall Stiffness Assessment Acoustic signals can also be used to estimate remaining effective AC pipe wall thickness. An acoustic signal is transmitted through the pipe and measured to determine effective (nondeteriorated) wall thickness. Although this technology cannot be used to identify specific locations of severe degradation, it can be used to estimate the average minimum effective remaining wall thickness over the length the acoustic signal is measured (typically a couple of hundred feet). This technology can be performed by affixing acoustic sensors to the exterior of the pipe or by inserting equipment into the pipeline with a cable.

Petrographic Analysis Petrographic examination of the pipe may reveal sulfate-induced deterioration and acid attack on the pipeline wall in addition to other signs of failure.

Note: The following rehabilitation and replacement sections are provided for information purposes only. They are not a component of the condition assessment process but are integral to an asset management program, and they were therefore included for reference.

9.4 REHABILITATION PRACTICES AND REPAIR MECHANISMS

Asbestos cement pipe may be considered hazardous material once it is cut and becomes friable. While working with AC pipe, appropriate safety regulations and measures should be followed. For example, keeping the AC pipe wet before and during cutting with soil pipe cutters can prevent the release of airborne asbestos fibers. In addition, once the pipe has been removed from a trench, the ends of the broken pipe, cut pipe, or any damaged areas should be immediately painted to seal any potentially friable asbestos fiber. Lastly, the pipe pieces (less than 3 ft) should be double bagged before transporting. [See AWWA Opflow Handling Asbestos-Cement Pipe: A New Approach (Fullen 2012a), for an example of safe handling work practices].

9.4.1 Internal Surfaces

Slip Lining Slip lining is a process in which a fabric liner is installed in an existing pipeline to address failure modes of leaking joints and/or compromised pipe walls. This technology has proven successful for AC pipe. Two types of liners should be considered: one that is structurally independent of the host pipe, considering internal and external loading, and one that is reliant on the host pipe to withstand internal and external loading. Slip lining does, however, marginally reduce the effective pipe cross-sectional area and hydraulic capacity.

Pipe Bursting Pipe bursting is a *trenchless* method of replacing buried pipelines (such as sewer, water, or natural gas pipes) without the need for a traditional construction trench. *Launching* and *receiving pits* replace the trench needed by conventional pipe laying. Successful pipe bursting of AC pipe has been demonstrated; however, once AC pipe is burst it becomes a hazardous material. Many jurisdictions consider sites where AC pipe has been burst as hazardous waste sites. Knowledge of EPA and local regulations regarding AC pipe handling and disposal is prudent when deciding whether pipe bursting is a good option. During the writing of this manual, pipe bursting of AC pipe was being reviewed by the EPA.

9.4.2 Exterior Repair Types

Clamps Repair clamps are typically made of stainless steel, are of *full circle* design with a rubber liner, and are readily available for all diameters of AC. A single full circle repair clamp can be used by itself for repairing circumferential breaks, replacing a failed coupling, and repairing holes that have been tapped for service line connection or even holes that have been unintentionally drilled or punched through the pipe wall. When installing a repair clamp over a circumferential break or when replacing a failed AC coupling, both ends of the pipe must be in alignment with one another. When a repair clamp is used to repair damaged pipe that has occurred from an auger bit or a boring tool, or simply to cover a service tap hole after the service saddle has been removed, the hole must not have any fractures extending from the borehole itself.

Couplings Repair couplings, often referred to as flex couplings, are made of ductile iron and have a ring and rubber gasket on each end that when tightened against the end of the coupling barrel compresses a rubber gasket that seals around the surface of the host pipe. These couplings are typically used when replacing a damaged section of AC and are used to join the ends of a new piece of pipe to the existing pipe. These same couplings are often used in a planned mode for pipe modifications.

REFERENCES

ASTM. (2006). "Standard test methods for asbestos cement pipe." *ASTM C500-98*, West Conshohocken, PA.

ASTM. (2011). "Standard test methods for asbestos-cement pipe." *ASTM C500/C500M*, West Conshohocken, PA.

ASTM. (2012). "Standard test methods for rubber property—Compression set." *ASTM D1415*, West Conshohocken, PA.

ASTM. (2016). "Standard test methods for rubber property—Compression set." *ASTM D395*, West Conshohocken, PA.

CHAPTER 10
PIPE INSPECTION TOOLS

Over the past decade, utilities have implemented asset management strategies that include condition assessment of pipelines. Condition assessment uses inspection data to assess the physical condition of an asset and provide an estimate of the remaining service life. When determining which inspection tool to use, risk is the most important factor to consider. For assets with a high consequence of failure, a detailed assessment is needed while lower risk assets may warrant a lower resolution screening technique.

Prior to implementing a condition assessment project or program, it is important to understand the anticipated failure modes for a given pipe material. Many of the tools that are currently available have been developed for specific pipe materials and detect defects associated with specific failure modes. As such, all the tools discussed in this chapter have limitations that should be clearly defined. Understanding the strengths and limitations of the tools and techniques will provide realistic expectations and improved understanding of the results.

This chapter provides an overview of several commonly used inspection tools and techniques that are currently available in the market. Due to the evolving nature of pipeline assessment, it is not practical to cover all of the tools or techniques that have been implemented. Rather, this chapter provides a sample of the more commonly used tools and techniques along with examples of the anticipated results.

10.1 INSPECTION TECHNIQUES FOR LOW RISK PIPELINES

Low risk pipelines, by definition, do not have a high consequence of failure or high likelihood of failure. As a result, significant investment in pipeline inspection is not always warranted. The inspection tools and techniques discussed in this section typically require lower financial or operational investment, are characterized as screening methods, and provide low resolution results.

10.1.1 Corrosion Surveys

- Material types: all
- Diameters: all

Corrosion surveys can be applied to all ferrous and concrete pipeline materials. They review a number of variables to evaluate the aggressivity of the environment surrounding the pipeline. These variables indicate that corrosion may be occurring on a pipeline. Some of the variables include

- Soil pH,
- Chloride content,
- Sulfate content,
- Resistivity,
- Drainage (wetting and drying),
- Redox potential, and
- Microbiological activity.

The American Water Works Association (AWWA) Standard C105 provides a 10-point scale to determine whether corrosion protection is necessary for cast iron (CI) or ductile iron (DI), based on soil corrosivity. Table 10-1 denotes the parameters scored under the system presented in AWWA C105. If the sum of the scores based on measurement of resistivity, soil pH, redox potential, sulfide content, and soil moisture equals 10 or more, then corrosion protection is recommended.

Tools and Techniques Available for Corrosion Surveys Corrosivity can be measured from the ground surface, above the pipeline alignment. A corrosion survey can detect and locate potentially corrosive environments surrounding the pipeline, providing a screening technique for the pipeline's condition. It should be noted that corrosivity surveys are low cost methods of identifying locations where corrosion *may* be occurring on the pipeline. They do not provide definitive results. A survey may miss areas of corrosion (i.e., false negatives) or may indicate corrosion when it is not occurring (i.e., false positives). This method of inspection should be viewed as a low cost, low resolution screening method, providing data that allow for an assessment of the pipeline's host environment.

Various survey methods include

- Pipe-to-soil potential,
- Cell-to-cell potential,
- Soil resistivity,

- Chemical analysis, and
- Linear polar resistivity.

Positives of Corrosion Surveys

- Low cost,
- Basic information on the environment surrounding a pipeline is provided,

Table 10-1. AWWA's 10-Point Scale for Soil Corrosivity

Soil Parameter	Score
Resistivity (ohm cm)	
<700	10
700–1,000	8
1,000–1,200	5
1,200–1,500	2
1,500–2,000	1
>2,000	0
Soil pH	
0–2	5
2–4	3
4–6.5	0
6.5–7.5	0
7.5–8.5	0
>8.5	3
Redox Potential (mV)	
>100	0
50–100	3.5
0–50	4
<0	5
Sulfide Content	
Positive	3.5
Trace	2
Negative	0
Soil Moisture	
Poor Drainage—Continuously wet	2
Fair Drainage—Generally moist	1
Good Drainage—Generally dry	0

Source: *Microcrystalline Wax and Petrolatum Tape Coating Systems for Steel Water Pipe and Fittings,* ANSI/AWWA C217.

- Very little to no pipeline modifications are required, and
- Pipeline operations usually remain unaffected.

Negatives of Corrosion Surveys

- Identifies only possible locations of corrosion,
- Does not provide quantifiable data on the actual condition of the pipeline, and
- Does not detect other pipeline deterioration mechanisms (e.g., leaks, third-party damage, thrust problems, internal corrosion, hydrogen embrittlement, or settlement issues).

Refer to the guidelines established by the National Association of Corrosion Engineers (NACE) RP0169 for detailed information on corrosion surveys.

Test pit locations may be selected to assess the actual condition of the pipeline and validate the results from any inspection tool or technique. Test pit locations can be chosen via both biased and random selection methods and may include locations that exhibited high corrosion potential during an external corrosion survey. The locations for excavation and inspection should be a representative collection of the overall site condition for the pipeline.

10.1.2 Closed-Circuit Television

- Material types: all
- Diameters: 8 in. and above

Closed-circuit television (CCTV) is a versatile tool that has been used for many years in pipeline management. While the development of standards for pressurized pipe has been attempted, CCTV inspection should be considered a screening method.

CCTV inspections require one access point for insertion and extraction of the tool, which is deployed both upstream and downstream of the access point (typically 1,500 ft can be inspected in each direction). It is important to note that multiple entry points along the pipeline will be required due to the limited length of the tether. The number of entry points is dependent upon the length of pipeline to be inspected.

Tools and Techniques Available for CCTV

Robotic CCTV Equipment. CCTV may be deployed in a pipeline using a robotic crawler or a tethered in-line probe. For example, the RedZone SOLO robotic crawler, shown in Fig. 10-1, is tethered only for retraction purposes. Different video resolutions are available using pan and tilt cameras. Panoramic cameras and digital postprocessing of images are also available.

Fig. 10-1. CCTV instrument
Source: Courtesy of RedZone Robotics

Fig. 10-2. CCTV instrument
Source: Courtesy of RapidView IBAK North America

A tethered, robotic vehicle performs the entire inspection in an out-of-service pipeline. The vehicle has a camera equipped with a pan, tilt, and zoom head, enabling a thorough view of the pipe wall in a variety of directions. Some equipment uses a fish-eye lens that allows the entire circumference of the pipe to be viewed without moving the camera. Some CCTV cameras are equipped with a sonde, either within the camera or tailored into a cable connector, allowing the camera's position to be tracked from above ground. Fig. 10-2 shows the IBAK Rapid View Pegasus, equipped with a high definition camera and LED lighting.

In-Service CCTV Equipment. A tethered probe is inserted through an access port into an operational pipeline. A camera is mounted on the probe

Fig. 10-3. CCTV Sahara tethered probe
Source: Courtesy of Pure Technologies U.S.

and the flow of water pulls the probe through the pipeline. The Pure Technologies Sahara video CCTV system, shown in Fig. 10-3, can be inserted into a live pipeline through a standard 2-in. or larger tap.

Positives of CCTV

- Identifies visually apparent defects such as failed linings, internal corrosion, excessive deformation, excessive settlement, and joint separation;
- Some units only require one access point for up to one mile of pipe inspection on all pipe diameters; and
- Allows visual defects to be identified and documented.

Negatives of CCTV

- May need the pipeline to be out of service (some untethered CCTV applications);
- Provides a limited assessment of the pipe wall, e.g., linings and coatings hide the wall, exterior corrosion is not evident, and a true structural assessment is not provided for pressure pipelines;
- Detects corrosion of metallic components but cannot quantify the extent of the damage;
- Must consider equipment pressure ratings, e.g., static head (out-of-service, but not dewatered, pipe);
- Limited by tether length (multiple insertions);
- Video quality for in-service CCTV can be affected by camera vibrations, bellies of water, and turbidity;

- Overlooks small defects due to video quality; and
- Cannot quantify severity of identified damage.

10.1.3 Sonar

- Material types: all
- Diameters: 8 in. and above

Sonar (short for sound navigation and ranging) is a technique for determining the ovality and diameter of a pipeline. A sonar pulse is generated onto the interior of a pipe wall to determine the dimensions of the pipe. This method is useful for identifying areas of a pipeline that may be out of round, areas that have experienced loss of the internal mortar lining, and areas where debris and sediment have built up in a pipe. Sonar can only be generated for sections of a pipeline that are full of water. Fig. 10-4 presents an example of the data output from a sonar inspection, which shows 0.46 ft of sediment on the invert of the pipe.

Sonar equipment can be attached to a robotic crawler to acquire sonar profiling data in conjunction with a CCTV inspection. Sonar can also be attached to a free-swimming, in-line inspection tool. Incorporating sonar

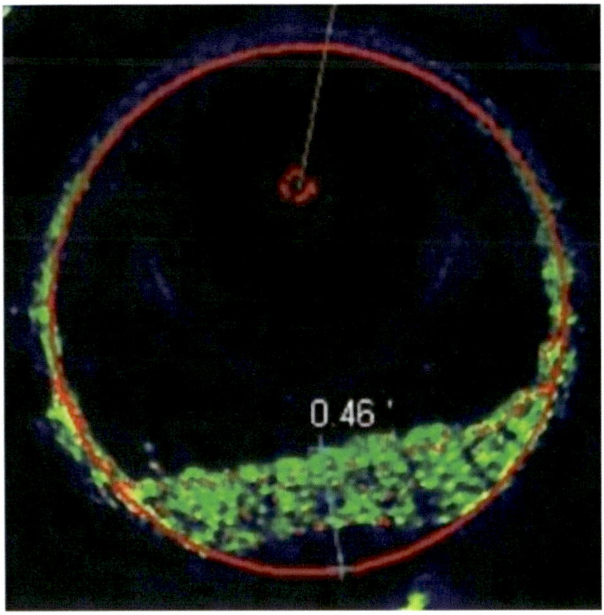

Fig. 10-4. Sonar data output showing debris at the invert
Source: Courtesy of Pure Technologies U.S.

into the pipeline assessment will augment the data by providing information for pipes that are full of water.

Positives of Sonar

- Determines the volume of debris in a pipe;
- Determines ovality in carefully controlled conditions;
- Excels as a navigation tool in zero visibility conditions, such as when the tool is under water in a gravity sewer or force main; and
- Can be deployed on a free-swimming platform for in-service pipelines.

Negatives of Sonar

- Can be slow if 100% coverage is desired;
- Takes 1.2 s or more per revolution with a 2-in. wide beam (traveling too fast results in helical data);
- Is difficult to align properly for good ovality measurements;
- Can provide misleading results to an untrained eye if debris and other factors are present;
- Has limited resolution;
- Cannot detect loss of coatings, or other wall loss, that is less than 5 mm (0.2 in.); and
- Can result in poor data due to fast flow and water bubbles.

10.1.4 Leak and Gas Pocket Detection Tools

- Material types: all
- Diameters: all

Pipeline leaks can be precursors to major failures for all materials. Ground shifts, poor installation, or rolled gaskets can lead to joint leakage. Once a leak forms, it can wash out the supporting bedding, which can in turn cause the pipe to shift and settle, further opening up the joint leak. This can, over time, lead to structural failure of the pipes. Leaks are also a primary source of nonrevenue water. A small leak will grow with time until it is identified and corrected.

Joint leaks can often be detected and repaired before causing structural failure. Leak detection on all forms of pressure pipe can be effectively performed by either in-line sensors (tethered or free swimming), or by using trunk main leakage correlators with hydrophones. In either case, operators with substantial experience in leak detection are required as this work is substantially different from leak detection on smaller diameter mains.

Air or other gas pockets can form in pressurized pipelines where high points in the pipeline profile are not properly vented with air release valves. In these locations, automatic air valves may be malfunctioning, which causes gas pockets to form. Gas pockets can also occur when high points exist where they were not designed, meaning that the high points are unvented. This situation can result from pipeline settlement or improper pipeline installation. Gas pockets contribute to a lower hydraulic capacity, compounding the effect of hydraulic transients, and may lead to internal corrosion in ferrous pipes.

Tools and Techniques Available for Acoustic Leak Detection

Acoustic Correlators. Acoustic correlators work by placing two acoustic sensors on a pressurized pipeline, separated by a known distance. By monitoring acoustic activity at these points, a leak can be identified on the pipeline between the sensors. The acoustic signals can be further evaluated to determine the location of the leak. Correlators cannot identify gas pockets.

To perform leak detection using correlators, access must be provided to the pipeline or to appurtenances that are acoustically coupled with the pipeline. Fig. 10-5 shows the TriCorr Touch correlators, which can be used on any pipe material and diameter.

Positives of Correlators

- Little to no disruption in flow.

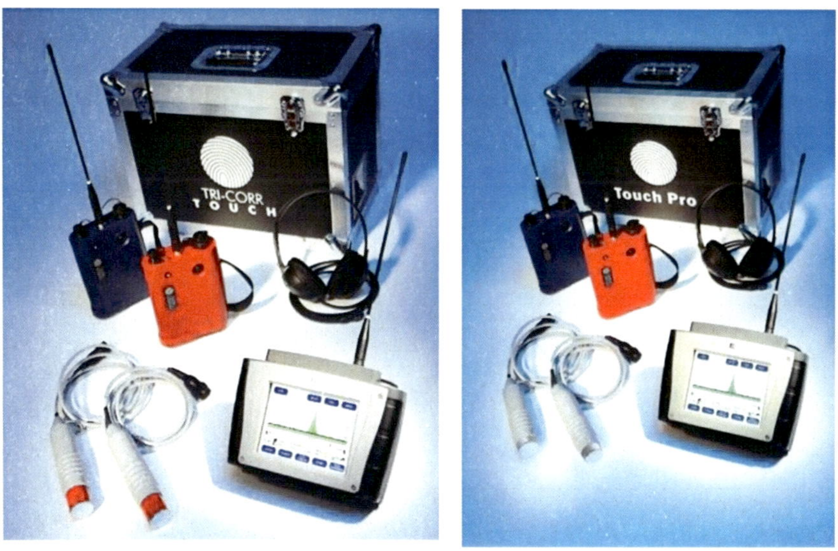

Fig. 10-5. Correlator by TriCorr Touch
Source: Courtesy of Fluid Conservation Systems

Negatives of Correlators

- Not as effective in large diameter pipelines, i.e., research has been performed on large-diameter pipelines but limitations persist;
- Multiple leaks between sensors cannot be differentiated (this limitation primarily affects large-diameter pipelines, where access points may be farther apart); and
- Gas pockets prohibit the use of correlators, i.e., gas pockets between sensors disrupt the propagation of the acoustic signal.

In-Line Acoustic Probes. This method involves inserting a probe into a pressurized pipeline. The probe then traverses the pipeline under normal operating conditions. Two configurations for these probes are available. Both configurations, which are detailed below, use acoustic sensors to detect the noise produced by leaks or trapped air within the pressurized pipe system.

Tethered, In-Line Acoustic Probe. A tethered configuration is one in which the probe is either attached to a push rod cable or a neutrally buoyant cable that is pulled by the flow of water down the pipeline. Sufficient flow velocity must be generated during the inspection to pull the probe down the pipe. Operators evaluate both visual (through CCTV) and acoustic data as the probe traverses the pipeline and then extract the probe once the inspection is complete. Data can be viewed live or recorded, and the instrumentation allows for operation while the pipeline remains in service. An access port must be provided to insert and extract the equipment from the pipeline. Access points usually consist of 2-in. or larger full port valves tapped on top of the pipe and fire hydrants can be used in some situations. Systems can be inclusive of technology that can measure internal/external pitting and current thickness of steel, cast iron, and DI pipes (Fig. 10-6).

Free-Swimming, In-Line Acoustic Probe. In this configuration a free-swimming device is inserted and allowed to traverse the pipeline with the flow of water. An example of a free-swimming device is shown in Fig. 10-7. At a point downstream, the device is captured and extracted from the pipeline. The acoustic data are then evaluated. For free-swimming tools, insertion and capture points must be carefully planned. Most devices can be inserted and captured through a 4-in. or larger valve tapped on top of the pipeline. Alternatively, insertion of the device can often be accomplished through check valves, temporarily removing the pipeline from service and opening a manhole, or through bypass piping. Alternative locations to capture a free-swimming device often involve building a customized screen or net that can be used at points of discharge to capture the tool.

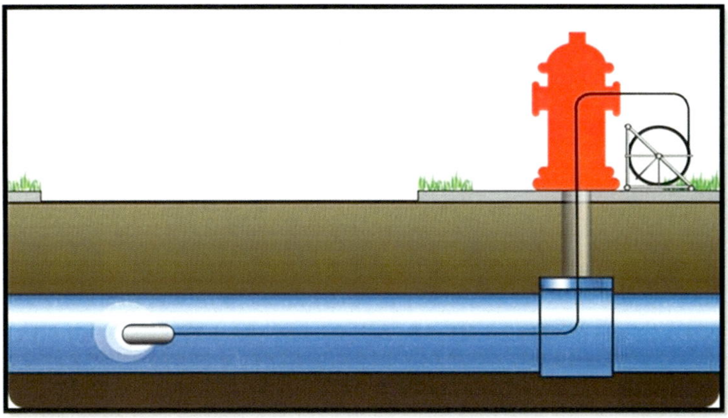

Fig. 10-6. Tethered probe
Source: Courtesy of JD7 USA

Fig. 10-7. SmartBall leak and gas pocket detection system
Source: Courtesy of Pure Technologies U.S.

Positives of In-line Acoustic Probes

- Applicable for all pipe materials;
- High sensitivity:
  - Can detect small leaks and gas pockets; and

   - Because the acoustic sensor travels in the pipeline, sensitivity of the device increases the accuracy of reported leak and gas pocket locations;
- Pipeline remains in service;
- Tethered probes can map leaks in real time;
- Free-swimming devices inspect long distances in short time frames;
- Minimal disruption or modifications to the pipeline; and
- Only two access points needed for free-swimming tools: insertion and extraction.

Negatives of In-Line Acoustic Probes

- Often requires certain pipeline configurations (e.g., closing of offtakes);
- Potential need for flow modifications;
- Free-swimming probe cannot be manipulated once inserted (flow must be controlled to ensure that the probe is not diverted);
- Minimum pressure requirements, ensures the correct pressure differential for the system to detect a leak;
- Ambient noise limits ability to detect leaks; and
- Possibility of losing tool or tool becoming trapped.

10.2 INSPECTION TECHNIQUES FOR MEDIUM RISK PIPELINES

Medium risk pipelines have a higher consequence of failure or likelihood of failure than low risk pipelines and warrant additional condition assessment investments. The tools described in this section could be used on their own or in combination with the methods described for low risk pipelines.

10.2.1 Acoustic Pipe Wall Stiffness Assessment

- Material types: CI, PVC, and asbestos cement (AC) pipe
- Diameters: 6 in. and above

Research by the National Research Council Canada has shown that the speed of sound through the water column in a pipeline is proportional to the stiffness of the pipe wall. Therefore, by measuring speed of sound in a pipeline, locations with reduced stiffness associated with pipeline deterioration may be identified. Technologies using this principle to identify deteriorated pipes, at the time of this writing, are relatively new with little independent verification.

Tools and Techniques Available for Acoustic Pipe Wall Stiffness Assessment

External Acoustic Pipe Wall Stiffness Assessment. This technique uses similar equipment to the correlators previously described, whereby acoustic sensors are affixed to a pipeline at a specific distance apart. Acoustic signals are introduced into a pipeline through hydrants or an acoustic pulser attached to the pipe surface at varying appurtenances. The acoustic propagation velocity is measured and inputted into an equation to calculate the minimum average pipe wall stiffness between the acoustic sensors. Depending on the pipe material, stiffness can be translated into the minimum average pipe wall thickness. For this tool, access must be provided to the pipeline or to appurtenances that are acoustically coupled with the pipeline. Fig. 10-8 presents the LeakFinderRT by Echologics, which is used to locate leaks through its acoustic pipe wall stiffness assessment capabilities.

Positives of External Acoustic Pipe Wall Stiffness Assessment Tools

- Noninvasive and no effect on pipeline operation;
- No pipeline modifications are required, may require excavation;
- Low cost; and
- Integrated with leak detection.

Negatives of External Acoustic Pipe Wall Stiffness Assessment Tools

- Measures the minimum average wall stiffness between sensors (no specific locations of defects);
- Technology is not independently confirmed (full capabilities and limitations are not understood);
- Gas pockets between sensors disrupt the acoustic signal, i.e., will not work when gas pockets are present;

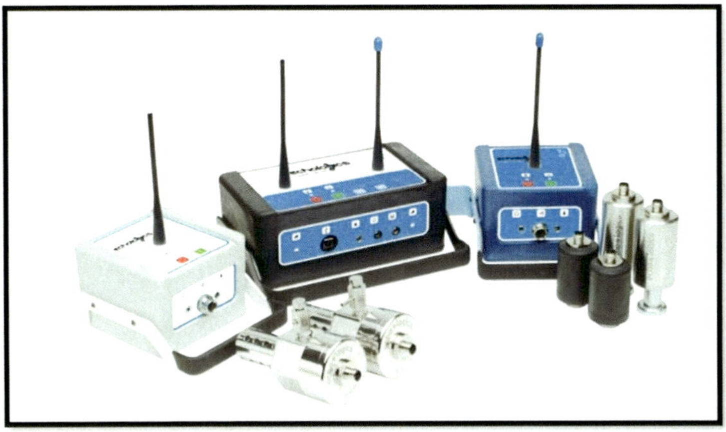

Fig. 10-8. LeakFinderRT external acoustic pipe wall stiffness assessment tool
Source: Courtesy of Echologics

- Additional access points along the pipeline may need to be provided through excavation; and
- Pressure requirements (minimum 15 psi).

Internal Acoustic Pipe Wall Stiffness Assessment. This technique relies on inserting a tethered probe into the pipeline. The equipment consists of an acoustic pulser and receiver, separated by a designated distance, that is calibrated for the pipeline. The pulser emits an acoustic wave at a designated time, and the receiver is used to calculate the speed of sound between the pulser and receiver. The average pipe wall stiffness between the sensors is then calculated. These measurements are recorded as the sensors move down the pipeline with the flow. An access port must be provided to insert the equipment into the pipeline. Access points usually consist of 4-in. or larger valves tapped on top of the pipe. Sufficient flow velocity must be generated during the inspection to pull the probe down the pipeline.

Positives of Internal Acoustic Pipe Wall Stiffness Assessment Tools

- Pipeline remains in service,
- Tethered probes can map deterioration in real time,
- Performed in conjunction with a leak detection survey, and
- Minimal disruption or modifications to the pipeline.

Negatives of Internal Acoustic Pipe Wall Stiffness Assessment Tools

- Often requires certain pipeline configurations (e.g., closing of offtakes),
- Potential need for flow modifications, and
- Inspection distance is limited by tether length (may require multiple insertions).

10.2.2 Handheld Eddy Current Scanning Method

- Material types: ferrous pipe
- Diameters: all

Handheld eddy current scanning methodologies measure the remaining average pipe wall thickness from either the interior or exterior of the pipe without removing the pipe lining or coating. In most cases, a grid consisting of longitudinal and circumferential lines is defined around the circumference of the pipe. The tool is moved to collect measurements at each grid point, establishing a gradient of wall thickness measurements around the pipe.

Tools and Techniques Available for Handheld Eddy Current Scanning Method The tools for this inspection method consist of an antenna, data cables, and acquisition equipment. The antenna typically covers an area between 1 and 4 $in.^2$ The acquisition equipment records data from the sensors, which is later evaluated to provide average pipe wall thickness. Fig. 10-9 shows an example of a handheld eddy current scanning system.

External scanning requires excavation of buried pipes while internal scanning can be carried out through manned entry of the pipeline. Some manufacturers offer tool configurations that allow for potholing. In this configuration, a group of sensors are mounted on a pole and lowered into the pipeline.

Positives of Handheld Eddy Current Scanning Method

- Provides average wall thickness for all types of ferrous pipe and detects wall thinning due to corrosion; and

Fig. 10-9. Handheld eddy current scanning
Source: Courtesy of Pure Technologies U.S.

- Does not require contact with ferrous pipe surface:
  - No modifications to coatings, linings, or surface corrosion (up to 2-in. thick) are required; and
  - Reduces the amount of pipeline preparation, compared with other evaluation methods.

Negatives of Handheld Eddy Current Scanning Method

- Requires excavation or manned entry;
- Provides relative pipe wall measurements only:
  - To determine the pipe wall thickness and subsequent remaining useful life, either assumptions must be made about the standard wall thickness or coupons must be collected to perform calibration measurements; and
  - Ultrasonic testing may also be used to collect actual wall thicknesses for DI and steel pipe;
- Does not detect cracks;
- Assesses the pipe wall condition only within the sample area:
  - Test pitting exposes a small percentage of the pipeline; and
  - Because pipelines do not degrade in a uniform fashion, data from such a small sample size can be misleading; and
- Provides the mean value of wall thickness for the area under the antenna (e.g., small but deep corrosion pits may not be detected if they are not within the tolerance of the tool).

10.2.3 Ultrasonic Testing Measurements

- Material types: metallic
- Diameters: all

Ultrasonic testing (UT) is a commonly used method to measure metallic pipe wall thickness and can be used on the exterior surface of a pipe while the pipeline remains in service or on the interior pipe surface when the pipeline is removed from service. A handheld instrument measures the propagation time of high frequency, short wavelength, mechanical waves through the base metal of the pipe wall. Defects are detected when the ultrasonic pulse is reflected at an interface between materials with different properties (e.g., graphite or cement mortar lining).

Tools and Techniques Available for UT The UT transducer (Fig. 10-10) is typically separated from the test surface (i.e., the pipe wall) by a couplant (such as oil or water). There are two methods of receiving the ultrasound waveform, reflection (pulse-echo) and attenuation. The attenuation method sends a pulse through the test sample to the transducer. In the reflection method, the transducer acts as both the sending and the receiving

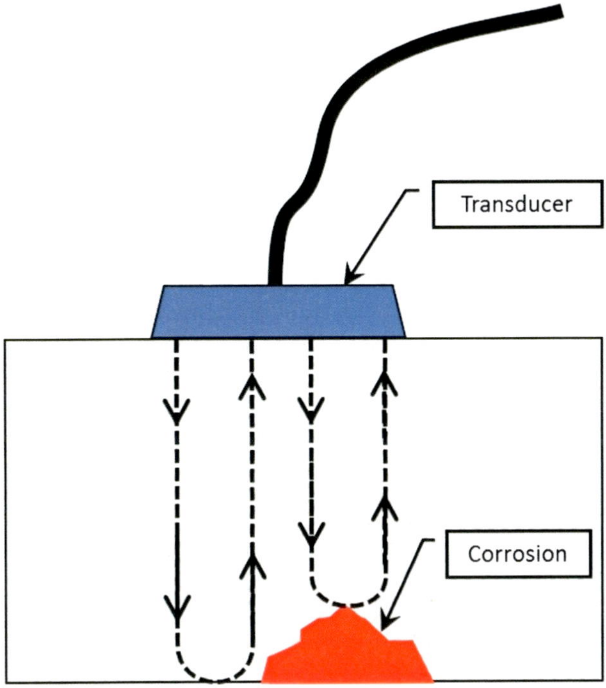

Fig. 10-10. Ultrasonic testing
Source: Courtesy of Pure Technologies U.S.

mechanism. Reflected ultrasound occurs at an interface, such as the back wall of an object or the face of a crack. The amplitude and arrival time of the pulse allows imperfections to be detected and quantified.

There are numerous ultrasonic tools available on the market.

Positives of UT

- Provides wall thickness measurements for ferrous pipes at discrete points,
- Widely used and well established, and
- Economical.

Negatives of UT

- Metallic surface must be clean (e.g., removal of any soil or corrosion, grinder recommended);
- Due to the properties of CI (consistent with the document) and DI (coarse grained materials), a lower frequency must be used with the UT tool than that typically implemented for steel;

- Obtains pipe wall measurements in sample area only; and
- Requires excavation or manned entry except where the pipe is exposed (e.g., valve vaults, air release valves).

10.2.4 Impact Echo

- Material types: concrete pipe
- Diameters: all

Impact echo is a nondestructive testing procedure that uses the principles of stress wave analysis to provide data about the condition of reinforced concrete structures, including pipelines. Stress wave analysis was developed during the 1980s, primarily to determine the strength and thickness of concrete members used for a variety of structures (bridges, buildings, and pipelines). The ability to measure thickness is of particular interest because delaminated areas have a reduction in thickness. As a result, impact echo is regularly used to identify delaminated areas of concrete members.

Impact echo technology uses the fact that a stress (sound) wave can be generated by striking the surface of the concrete with a mechanical device. This wave propagates through the structure and is reflected by internal flaws or external surfaces. The speed of propagation may also be related to the amount of prestress in the concrete. A low tech deployment of impact echo will have a person listening for variations in the sound of a "hammer strike" on the inner surface of a concrete pressure pipe, typically using an impulse hammer like the one shown in Fig. 10-11. More sophisticated deployments include the regulation of the strike force using a Schmidt trigger or a controlled shot from a steel projectile, and the recording of the sound using piezoelectric transducers.

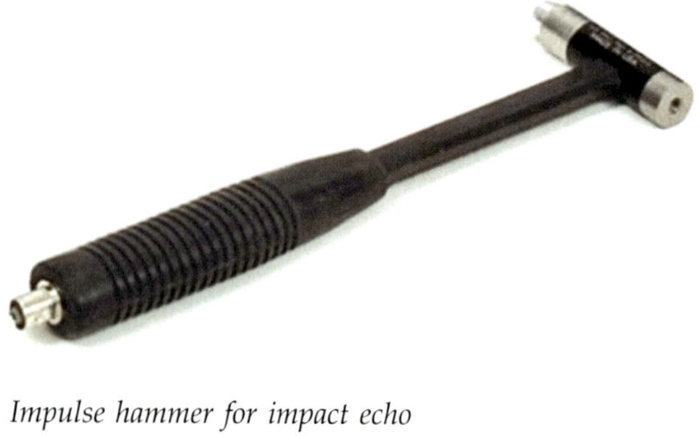

Fig. 10-11. Impulse hammer for impact echo
Source: Courtesy of Olson Engineering

Tools and Techniques Available for Impact Echo There are multiple vendors of impact echo technologies, each with its own capabilities and limitations. Some technologies consist of single point transducers and receivers. At least one manufacturer uses multiple receiving points to provide a better understanding of the structure.

Impact echo technologies have been used on both the interior and exterior surfaces of concrete pipelines, and each deployment has certain operational requirements. For internal inspections, the pipe needs to be large enough to allow for manned entry and the line must be dewatered. For external inspection, there is no size constraint; however, the line must be excavated.

Positives of Impact Echo

- Tests for indicators of reinforced and prestressed pipe wall condition, includes areas of delamination and areas of wall thinning;
- Used on the interior or exterior surface of a pipe; and
- Significant training is not required to collect measurements.

Negatives of Impact Echo

- Pipeline needs to be either dewatered or excavated, depending on the evaluation method;
- Proper surface preparation of the concrete is required;
- Only point measurements are collected (i.e., many measurements are typically needed for truly representative results);
- Expertise is needed to interpret the results; and
- Method is labor and time intensive.

10.3 INSPECTION TECHNIQUES FOR HIGH RISK PIPELINES

High risk pipelines have the highest consequence of failure or likelihood of failure and warrant comprehensive condition assessment investments. The tools described in this section could be used on their own or in combination with the methods described for low and medium risk pipelines.

10.3.1 Electromagnetic Inspection

- Material types: prestressed concrete cylinder PCCP
- Diameters: greater than 16 in.

Electromagnetic (EM) inspection tools are a nondestructive testing technique that can detect and locate broken prestressing wire wraps in PCCP. These technologies can identify structurally compromised PCCP before failures occur. EM inspection technologies use a transmit-receive eddy current system in which both the exciter and detector are located inside the pipe. Little to no axial separation between the exciter and the detector is required to ensure that the signal received by the detector is dominated by a through transmission effect. This is because the presence of the externally wound prestressing wire acts to amplify the magnetic field that passes through the pipe wall. The disruption of this amplification effect indicates broken prestressing wire wraps; hence, broken wire wraps are characterized by weaker signals with more time delay. Because many prestressed pipes also contain steel cylinders, EM inspection technologies may be impacted by changes in the cylinder or changes in other EM properties within the pipe.

Tools and Techniques Available for EM Inspection EM inspection tools induce an electrical field in the prestressing wire and then measure the corresponding EM signature. An exciter coil creates an electrical field that is amplified by the helically wrapped prestressed wire. A detector coil records the subsequent EM signal. PCCP with intact prestressing wire wraps produces consistent EM signals while a PCCP with broken prestressing wire wraps causes distortions in the EM signal, as shown in Fig. 10-12. Analysis of the distortion characteristics (length, magnitude, phase shift, etc.) indicate the quantity of broken wire wraps, which is a strong indicator of the structural integrity of the PCC pipe. The spike in the green line and dip in the red line indicate broken wire wraps.

EM inspection tools can be deployed internally using three techniques: manned, robotic, and free swimming. Fig. 10-13 shows a robotic inspection platform while Fig. 10-14 illustrates a free-swimming tool. Fig. 10-15 shows an external manned testing device.

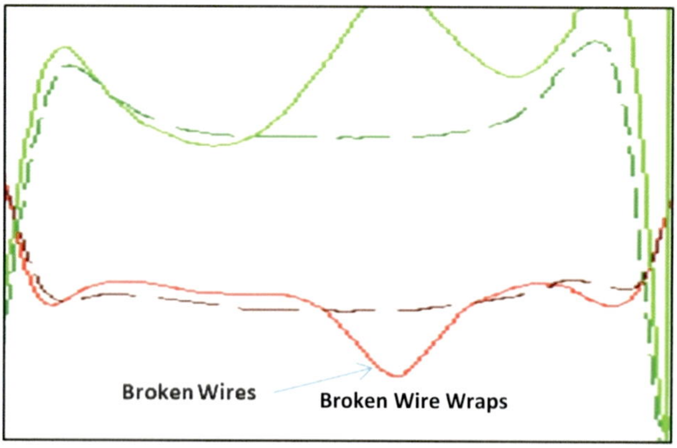

Fig. 10-12. Example of the electromagnetic signal for a damaged PCCP
Source: Courtesy of Pure Technologies U.S.

Fig. 10-13. PureRobotics platform robotic electromagnetic inspection tool
Source: Courtesy of Pure Technologies U.S.

Fig. 10-14. PipeDiver free-swimming electromagnetic inspection tool
Source: Courtesy of Pure Technologies U.S.

Positives of Manned EM Inspection

- Most thorough and accurate nondestructive PCCP assessment, can be correlated to internal visual inspection results; and
- Straightforward reinspection, if necessary.

Fig. 10-15. PipeScanner external RFEC testing
Source: Courtesy of Pure Technologies U.S.

Negatives of Manned EM Inspection

- Pipeline must be shut down and dewatered; and
- Minimum pipeline diameters for internal inspections: 36 in. for water pipelines and 60 in. for wastewater pipelines, with varying requirements for entry port spacing and ventilation.

Positives of Robotic EM Inspection Tool

- Comprehensive PCCP assessment when combined with other sensors (e.g., CCTV),
- High definition visual inspection option, and
- Additional sensor deployments are feasible.

Negatives of Robotic EM Inspection Tool

- Pipeline must be depressurized,
- Minimum entry port diameter is 16 in.,
- Minimum pipeline diameter is 18 in., and
- Maximum deployment length of approximately 6,000 ft (average deployment is between 2,500 and 3,500 ft).

Positives of Free-Swimming EM Inspection Tool

- Provides accurate PCCP assessment,
- Conducted under full operating conditions, and
- Capable of long-range deployments.

Negatives of Free-Swimming EM Inspection Tool

- A minimum of 12-in. flanged adapters are necessary for insertion and extraction,
- Lower resolution data than manned or robotic inspections,
- Minimum pipeline diameter of 16 in. required if no 90° bends exist, and
- Minimum pipeline diameter of 24 in. required if 90° bends exist.

10.3.2 Eddy Current/Enhanced EM Inspection

- Material Types: ferrous pipes and bar wrapped pipe (C303)
- Diameters: greater than 18 in.

Eddy current tools are a nondestructive testing method that is based on the principle of impedance change in an EM or exciter coil as it is scanned over a conductive surface (ferromagnetic material). Eddy current testing uses EM induction to detect flaws in conductive materials. A time varying current is passed through an electrical coil that generates a magnetic field. When the coil is placed near a conducting test object, eddy currents are generated in the test object in opposition to the magnetic field. Variations in the amplitude and phase of these eddy currents can be used to detect defects and material changes in the test object.

Eddy currents are induced as an energized coil is brought near the surface of a metallic pipe or metallic component of a pipe. The existence of flaws or variations in material properties causes distortions in these eddy currents, which alter the impedance in the exciter coil. This change in impedance allows for interpretation of the type of flaw or condition change.

It is crucial to note that cracks perpendicular to the path of the current interrupt the eddy current flow, allowing for detection. Cracks parallel to the flow path do not cause significant interruption and thus may not be detected.

Tools and Techniques Available for Eddy Current Distinct eddy current testing methods can be derived from the manner in which the eddy currents are detected and analyzed. These methods are presented below.

Impedance Probes. When a single frequency current is passed through a coil of wire, the coil will have certain impedance when it is far away from any test object. Next to a test object, this impedance will change (due to the presence of eddy currents in the object). Variations in the electrical conductivity and magnetic permeability of the test object, or the presence of any flaws or cracks will cause a change in the induced eddy currents, altering the impedance of the test probes. Thus, by maintaining a constant current within the probe and monitoring the voltage, changes in the amplitude and phase of the impedance can be monitored and information about the test specimen can be obtained. In general, impedance probes behave poorly when not in direct contact with the test specimen (e.g., testing in a mortar lined steel pipeline).

Pulsed Eddy Current Probes. Multiple frequencies or even a current pulse can be used to energize an eddy current probe. In general, the higher the frequency of the current, the less "depth of penetration" will be achieved in the test object. Using a pulsed eddy current can allow the thickness of the pipe wall to be estimated and, generally, can overcome larger lift-offs from the test surface than single current probes.

Transmit-Receive Eddy Current Probes. In transmit-receive probes, both a transmitter and receiver coil are used. Similar to impedance probes, the transmitter generates a magnetic field that induces eddy currents into a test object. Often, the receiver is on the opposite side of the test object (or surrounds it) and the voltage received in this coil provides information about the variations about material properties and geometry of the specimen.

RFEC. This technology detects both internal and external pipe defects with the use of multiple magnetic coils, an exciter coil, and one or more detector coils.

Two coupling paths exist between the exciter and detector coil(s), direct and indirect paths. The direct coupling path is generated inside the pipe and attenuates quickly due to circumferential eddy currents induced in the pipe wall. The indirect coupling path travels quickly along the pipe, encountering very little attenuation and exhibiting shifts in signal magnitude and phase when reaching any discontinuities (e.g., pipe defects) in the path.

RFEC is an example of a transmit-receive eddy current system that is used for the detection of flaws in ferromagnetic (steel, CI, and DI) pipelines. An RFEC probe consists of an exciter and detector that are both inside the pipeline being investigated but are spaced approximately two to three pipe diameters apart, depending on the operating frequency. At this spacing, the signal that reaches the detector is proportional to the magnetic field generated by the exciter as the field passes through and out of the pipe wall close to the exciter and passes into and back through the pipe wall close to the detector. As the magnetic field passes through the pipe wall, it is slowed down and becomes weaker. Areas of metal loss are characterized by increases in amplitude and decreases in phase delay, and both internal and external defects are detected with approximately equal sensitivity.

Fig. 10-16. In-line RFEC see snake tool
Source: Courtesy of PICA USA

In-line RFEC inspection tools can be used for the detection of ferrous pipe wall defects such as pitting, corrosion, cracks, and leaks, and can measure remaining wall thickness, surface area, and locations of pipe stress, although the amount of stress itself is not quantified. An example of an in-line RFEC tool is shown in Fig. 10-16.

Positives of RFEC Wall Assessment

- Identifies wall loss along the entire inspected pipeline,
- Higher resolution than acoustic pipe wall assessment tools,
- Industry standard for metallic pipe wall assessment, and
- Free-swimming platforms are capable of long-range deployments.

Negatives of RFEC Wall Assessment

- Available in-line tools are size on size, meaning that pipeline modifications are typically required for tool insertion and extraction (requires full bore for insertion and extraction).

10.3.3 Magnetic Flux Leakage

- Material types: ferrous
- Diameters: 4, 8, 12, and 24 in. (in-service); greater than 36 in. (manned)

Magnetic flux leakage (MFL) tools were initially developed for the oil and gas industry to inspect critical steel pipelines carrying product at extremely high pressures. Modifications to this technology have allowed for the development of tools capable of inspecting ferrous water pipelines with inner mortar linings of up to 1 in.

MFL is a method of nondestructive testing that creates a magnetic circuit in a ferromagnetic pipeline. Once the circuit is established, any defects in the test pieces will cause the magnetic flux to leak out of the pipe wall. This leakage is detectable using sensors placed between the poles of the permanent magnets that are on the tool. MFL may be used for steel, DI, and CI (all ferromagnetic) pipelines. The system is sensitive to localized pitting and general corrosion; however, complex regions of corrosion are more difficult to interpret than simple ones.

Tools and Techniques Available for MFL The basic principle of MFL is to magnetize the pipe wall with strong, rare-earth magnets and directly measure the magnetic field leakages caused by pitting or other defects. Fig. 10-17 shows a typical schematic diagram for a single MFL detector. Typical MFL tools place a detector every 0.25 in. along the circumference of the pipe, enabling extremely high resolution results.

The PureMFL tool, shown in Fig. 10-18, is a smart pig that indicates the outer diameter of the tool is only slightly less than the inner diameter of the pipeline that it inspects. Therefore, traversal of the pipeline could be inhibited by excessive ovality, sharp bends, or tuberculation.

External MFL systems may also be used to detect localized pitting or general corrosion on ferromagnetic pipes. An example is shown in Fig. 10-19.

In-Service Inspections. Pigs are generally added to and removed from an in-service pipeline at pig launching stations, a feature that is generally not included in water mains. Therefore, in-service MFL inspections often require

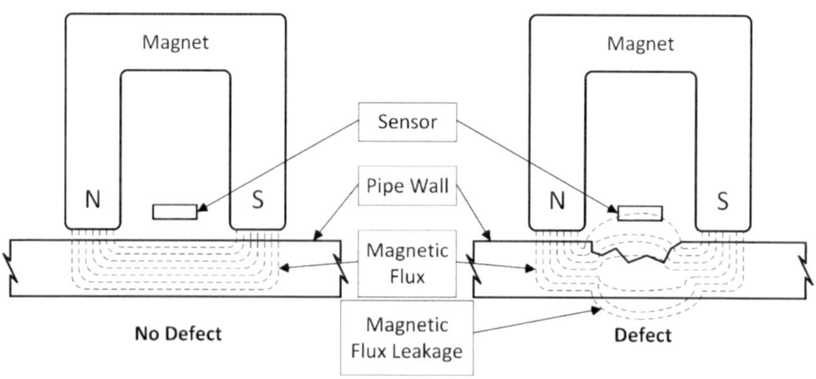

Fig. 10-17. Magnetic circuit
Source: Courtesy of Pure Technologies U.S.

Fig. 10-18. PureMFL inspection tool
Source: Courtesy of Pure Technologies U.S.

Fig. 10-19. External MFL tool
Source: Courtesy of Advanced Engineering Solutions Limited

modifications to the pipeline at the insertion and extraction points. Also, the pig can only pass through in-line valves that offer a full diameter open port, so inspections are typically carried out between valves. Before the inspection can be conducted, cleaning pigs should be used to ensure that the pipe walls are free of debris and that there are no bends or deflections too great for the tool to pass. Once the pipeline is cleaned, the MFL tool is used for inspection.

While in-service MFL tools can be free swimming, it is advisable to keep the tool tethered so the exact inspection distance is known.

Dewatered Inspections. Dewatered inspections are generally performed on larger diameter pipelines (36 in. and greater). These inspections require manned entry, so proper ventilation and confined space safety is necessary. First, a profiling tool measures the shape of the inner diameter to ensure that the MFL tool will fit. Then, the MFL tool is assembled inside the pipeline. Fig. 10-20 shows an example of data output for an MFL inspection.

Positives of MFL Inspection

- High resolution data:
  - Inspects nearly the entire circumference of the pipe wall, allowing for comprehensive structural evaluations; and
  - Collects data through interior mortar linings up to 1 in. thick;
- Industry standard for metallic pipe wall assessment; and
- Manned or free-swimming deployment options.

Negatives of MFL Inspection

- Access is required for insertion and extraction of the full diameter tool;
- High cost;

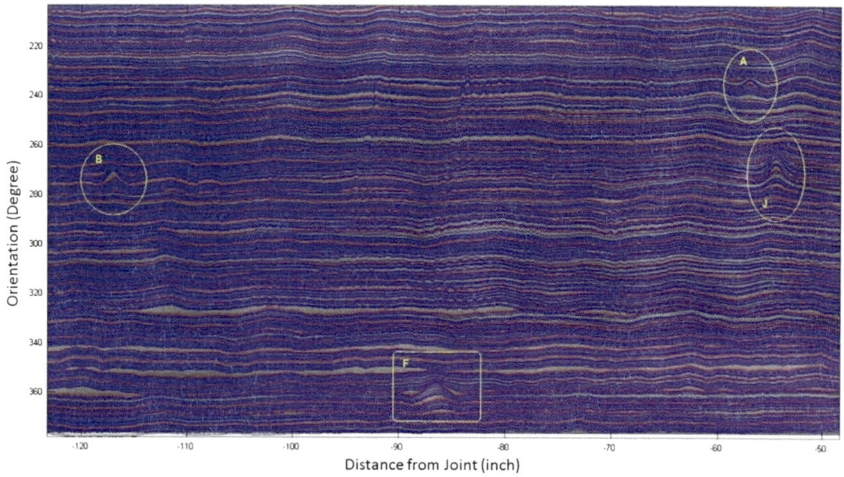

Fig. 10-20. MFL data output example, with defects circled
Source: Courtesy of Pure Technologies U.S.

- Requires a calibration tool;
- Limited by pipeline bends and internal features, e.g., valves; and
- Minimum flow requirements to propel the free-swimming tool.

10.4. RECENT DEVELOPMENTS IN CONDITION ASSESSMENT TOOLS AND TECHNOLOGIES

Pipeline condition assessment is a rapidly growing field, and tools and techniques are routinely introduced into the industry. Recently, new tools and techniques for the assessment of varying types of pipelines have been made commercially available. These tools have not yet been widely adopted in the industry and, as such, were not included in the main body of this chapter. These new tools are described below.

10.4.1 Pipe Wall Stress Assessment

- Materials: ferromagnetic pipes
- Diameters: 6 to 60 in.

Differences in the level of stress in a ferromagnetic material (CI, DI, and steel pipe) affect the material's response to an induced electrical current. As a metallic pipe wall corrodes, it loses cross-sectional area, increasing the level of stress in the metal. Recent advancements in the analysis of EM results have determined that in-line tools can detect variations in the pipe wall stress due to changes in the properties of the pipe wall (internal or external corrosion, changes in pipe material, etc.). Erosion of the bedding material, buckling, denting, wrinkling, shifting of a pipeline restraint (e.g., an anchor block), and unstable surrounding soils can also cause localized stress increases in the pipe wall. Once these areas are identified, external evaluation tools, such as handheld eddy current and UT measurements, can be used to determine the extent of the wall loss. In addition to potential areas of corrosion, pipe wall stress assessment tools can also detect material changes, pipe joints, and most appurtenances.

10.4.2 Acoustic Pipe Wall Stiffness

Tools that externally and internally measure the pipe wall stiffness acoustically were discussed in Section 10.2.1, "Acoustic Pipe Wall Stiffness Assessment." While these tools have been used in the industry for the inspection of CI, PVC, and AC pipes, recent enhancements to some of the tools may have made them applicable for the evaluation of PCC pipe.

The prestressing wire wraps in PCC supply strength to the pipe by holding the concrete core in compression. This compression increases the stiffness of the pipe in the hoop direction. Measuring the hoop stiffness of the pipe can provide a qualitative indication of loss of wall compression energy. Specialized acoustic technology may be used to measure the average hoop stiffness between test points through the acoustic velocity propagation measurement method (also known as acoustic pipe wall assessment). This method is used mainly as a screening tool for areas of potential concern or a prioritization tool for large amounts of pipe, as it does not indicate individual regions of degradation for spot repair and as such may not be applicable when making engineering decisions.

10.4.3 Continuous Leak Monitoring

- Materials: all
- Diameters: all (fiber optic monitoring); 12 to 120 in. (acoustic-based monitoring)

Monitoring transmission mains for the formation of new leaks presents an opportunity to repair leaks as they form, further reducing the chances of a leak developing into a full failure. Two types of products are commercially available for continuous leak monitoring: fiber optic lines buried nearby the pipeline that monitor for temperature changes and acoustic hydrophone-based trunk main monitoring devices.

Tools and Techniques for Continuous Monitoring

Fiber Optic, Temperature Change Leak Monitoring. This tool is implemented during new pipeline construction. Fiber optic cable is laid in the trench before the pipes are installed. The system monitors the surrounding soil for localized changes in temperature, which could indicate a leak in the pipeline.

Acoustic Leak Monitoring. Using hydrophones, a pipeline is continuously monitored for acoustic vibrations that indicate leaks and transient events. Sensors are mounted on the exterior surface of the pipeline at existing features, and data are sent to a secure server for analysis.

Acoustic PCCP Monitoring. Using fiber optic sensing cables, a PCC pipeline is continuously monitored for acoustic vibrations that indicate wire breaks along individual pipe sections. A permanent acoustic fiber optic (AFO) cable can be installed under normal operation or while the pipeline is dewatered for distances up to 25 km, per the data aquisition system. The system monitors the pipeline in near real time and reports the time and location of discrete wire breaks along the length of the pipeline. The information from the AFO system (rate of deterioration) can be used in combination with internal EM inspections (baseline condition) to establish the remaining useful life of individual pipe sections.

REFERENCES

AWWA (American Water Works Association). (2010). "Polyethylene encasement for ductile-iron pipe systems." *AWWA C105*, Denver.

AWWA. (2016). "Microcrystalline wax and petroleum tape coating systems for steel water pipe and fittings." *AWWA C217*, Denver.

NACE (National Association of Corrosion Engineers). (2013). "Control of external corrosion on underground or submerged metallic piping systems." *NACE RP0169*, Houston, TX.

CHAPTER 11
CONDITION ASSESSMENT REPORT

11.1 INTRODUCTION

The content of a condition assessment report will vary from project to project and utility to utility. Variations in report content are attributable to the individual needs of the project, type and size of pipe, and transmission versus distribution, as well as the budget constraints at the individual utility. This chapter, however, provides the utility with guidelines for a comprehensive condition assessment report that should help the utility make decisions on project prioritization and implementation. It is intended that the utility can consider these recommendations, either supplementing or eliminating work items to help form a scope of work for their project.

11.1.1 Life Cycle of Pipeline

When referring to pipeline life, it is important to clearly define the level of life being identified. The three primary definitions are design service life, effective service life, and the end of life, as shown in Fig. 11-1.

Design service life refers to the pipeline meeting the original design criteria with the original design safety factors. It should be understood, however, that some very old pipelines (e.g., gray cast iron pipelines that are a century old, or even older) may have been installed under vastly different design criteria, safety factors with regard to modern combined loading theory, or different load factors than what are used with rigid pipe today.

Effective service life is the time frame during which the original safety factors have been exceeded and the time when it is expected the pipeline may no longer function with structural integrity. At that latter point, the pipeline is considered to have reached its end of life because it will be

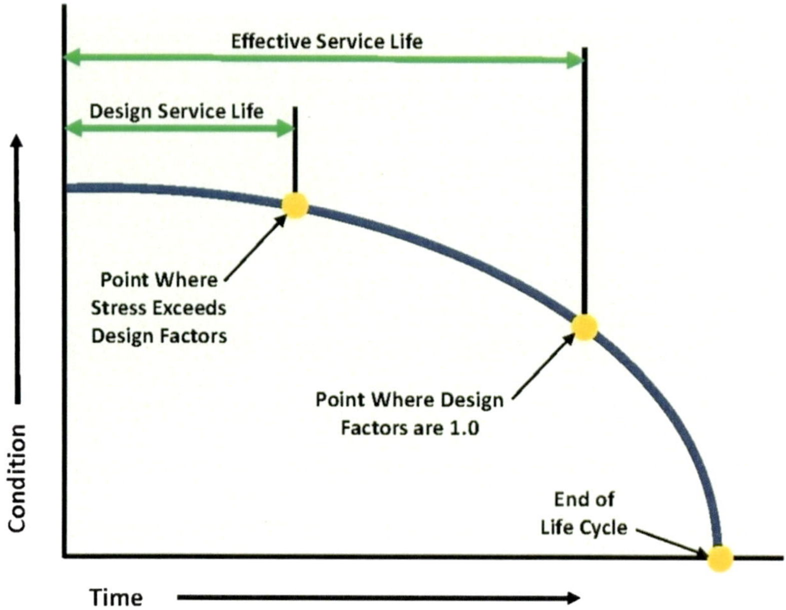

Fig. 11-1. Life cycle of a pipeline
Source: Courtesy of Carollo Engineers

very difficult to provide accurate predictions regarding the potential time frame for further life.

Each definition of pipeline life has purpose, benefit, and level of concern when incorporated in an asset management program.

11.1.2 Goal of Condition Assessment Program

The primary function of a successful condition assessment program is to identify in which of these three life periods the assets are presently; what, if anything, is affecting the expectation of each asset to remain in that life period; and, should some condition be affecting that asset, what would be a reasonable prediction of the remaining service life at that existing level. The program must then monitor those assets to, in effect, follow up and determine whether the predictions were and are reasonable. The process for implementing a condition assessment program is shown in Fig. 11-2.

It is important to recognize that condition assessment is strictly a process for identifying an issue at a point in time and providing a predictive evaluation of the rate that the asset is changing with time. The condition assessment process is one part of a good asset management program. The process of condition assessment can be completed with the use of one or many tools. Any effort to alleviate a condition adversely affecting an asset is a function of

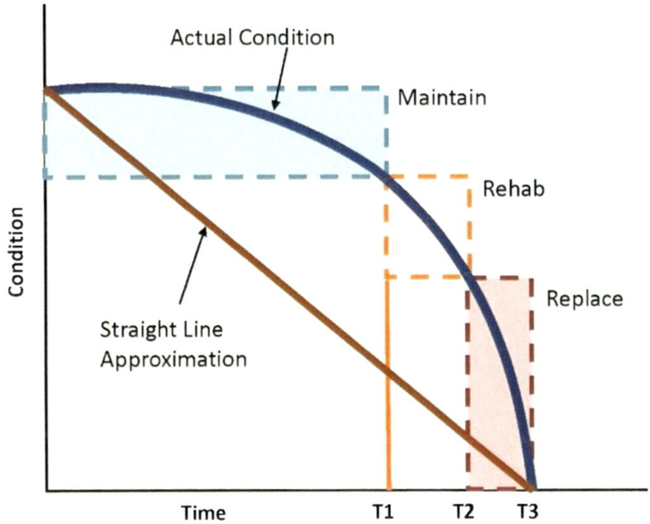

Fig. 11-2. Deterioration curve
Source: Courtesy of Carollo Engineers

asset management, not condition assessment. Assets in poor condition should be addressed via the utility's renewal and replacement program, also referred to as a capital improvement program (CIP), and/or the utility's operations and maintenance program, which may include CIP funding.

Test results from a condition assessment program will tend toward a definitive or comparative profile. Current assessment tools, many of which are highly effective at identifying and characterizing specific causative parameters, will rarely clearly define the remaining pipeline service life. Therefore, the managers of a condition assessment program should not be surprised if the results do not clearly define the exact condition of each asset but, instead, compare asset condition to other assets to gauge the level of deterioration and remaining life.

When a condition assessment has been made of an asset, particularly in a larger system, it will likely be necessary to apply some numerical value to the asset to define its level of concern in the system. Such an evaluation would again be individualistic and dependent on the nature of the utility and its resources.

11.2 SUMMARY OF FIELD TESTING PROGRAM

A good field investigation program forms the basis for the condition assessment. Any condition assessment report should include a comprehensive description and documentation of the process for which the data are acquired. This section provides recommendations for the documentation of the data acquisition process.

Documentation includes the following:

- Description of existing information available and reviewed for the project;
- Type of testing;
- Frequency of testing;
- Decision process for locating tests;
- Documentation of testing procedure, apparatus, and calibration;
- Verification;
- Material sampling and laboratory testing;
- Compilation and archiving of raw testing data; and
- Identification of future investigations.

11.3 DATA ANALYSIS

The information collected from a field program falls into two general categories: one that describes the pipe environment and one that determines the pipe condition. A totally comprehensive pipeline analysis would involve exposing the entire exterior of the pipe and examining with equal rigor the interior of the pipe. However, this is cost-prohibitive, not to mention logistically impractical. Instead of totally exposing the pipe and analyzing it, data from strategically placed investigations are used to predict the condition along the route. Data obtained for the testing program are then analyzed to determine the physical condition of the pipeline, assess the remaining service life of the pipeline, and make recommendations on how to proceed with the rehabilitation or replacement of the pipeline. This may include

- Examination of any available history or documentation regarding service and maintenance on the line, e.g., breaks, leaks, cost of any problems, any injuries to workers or the public, as well as other pertinent information;
- Evaluation of the pipe environment including soil corrosivity, pipeline external loading conditions, potential for changes to the loading in the future, and bedding condition;
- Evaluation of visual data obtained from the investigation including closed-circuit television; photographic evidence for condition of joints, linings, and coatings; and visual evidence of leakage or corrosion;
- Evaluation of physical data obtained during the investigation including wall thickness measurement for metallic pipelines, soundings, and evidence of wire breaks; and

- Evaluation of laboratory testing results for material properties of plastic and asbestos cement pipe and metallurgical properties of metallic pipe.

All the factors affecting the pipe environment and the pipe condition must be carefully observed and analyzed to assess not only the current condition of the pipeline but also factors that could affect its condition into the future. Fig. 11-3 illustrates the process by which this information feeds into subsequent analyses.

11.4 STRUCTURAL ANALYSIS

The tested pipeline must be analyzed to determine the ability of the pipeline to withstand the loads to which each section is exposed. The results of this analysis will determine where the pipeline is in its effective service life in comparison with its original or currently applicable design criteria.

The analysis will also determine which pipeline sections must be replaced or rehabilitated because they have exceeded their effective service life versus which pipes remain serviceable. Evaluating the pipelines in this way recognizes their ability to continue to be reliable even after some corrosion or deterioration has occurred.

The structural analysis performed here incorporates actual conditions observed during the data collection process and includes actual load conditions and current condition of the pipe. To assess the pipe's ability to serve its purpose into the future, this analysis must also take into consideration any future planned operational changes, potential for changes in external loading, and an assessment of future deterioration of the pipeline.

Segments with unacceptably low factors of safety are identified for incorporation into the recommended replacement or rehabilitation program. Prior to interpreting the pipe condition data, an understanding of the initial (new) condition of the pipe must be established. In this way, the existing condition of the pipe can be compared with the condition of the pipe when initially placed in service. Using this comparison, it can be determined at what stage the pipe is in the life cycle.

11.4.1 Methodology

The structural analysis for the pipeline consists of an evaluation of the following:

- Internal loads from operating, and transient pressures acting from within the pipe;

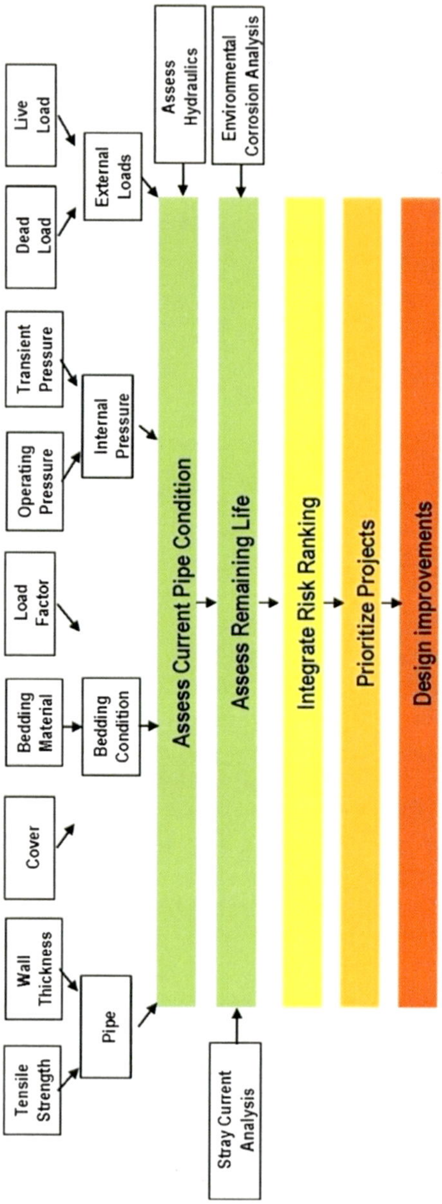

Fig. 11-3. Data analysis flowchart
Source: Courtesy of Carollo Engineers

- External loads from soil and structures (dead) and traffic (live) transmitted through the soil and acting upon the outside of the pipe; and
- Pipeline material properties.

Operating Pressure (p_o) Operating pressure loads are internal pressure loads that the pipeline is routinely exposed to during normal operation. The operating pressure load acting at a point on the pipeline is calculated from the difference between the hydraulic grade line and the elevation of the pipeline at the location in question.

Transient Pressure (p_t) Transient pressure loads are extraordinary loads resulting from hydraulic transients that can occur within the system. These transients can be caused by the closing of valves, pump starts and stops (e.g., in power outages), emergency shutdowns, excessive water velocity in filling of lines, and other events. Transient pressures must be determined individually for each pipeline.

Dead Loads (w_d) For the purpose of most analyses (excluding pipelines beneath heavy structures), dead load is the weight of soil above the pipeline. Where the pipe was exposed, the height of soil above the pipe is directly measured by field crews. At other areas, the height of soil is based on surveyed ground elevation and records on file that depict pipe elevations.

Live Loads (w_l) Live loads are vehicle loads (either from highways or railroads, depending on the facilities) located at the ground surface above the pipe.

Load Combinations A pipeline can be exposed to internal or external loads and a combination thereof.

Material Properties Material properties must be determined or estimated from existing information and historical data or laboratory testing of coupon samples taken from the pipe.

11.4.2 Estimation of Remaining Service Life

The prediction of remaining service life is required to evaluate future safe operating strategies including reinspection and maintenance schedule, as well as forecasting capital improvement for rehabilitation and/or replacement of a pipeline. As a result, it can minimize the possibility of pipeline failures until it reaches the end of its service life.

Assessment of the remaining service life of a pipeline with any high degree of certainty is difficult at best. It requires the use of structural analysis, statistical analysis, and good engineering judgment. Because of these difficulties, many methodologies focus solely on a ranking analysis approach in which the relative condition of each pipeline is compared, a recommendation of required remedial action is determined, and a prioritization of action is suggested.

Although this approach may be acceptable in many instances, assessment of a conservative service life is important in others. Different methodologies have been used, depending on pipeline material, to assess the remaining service life. Most methodologies focus on the assessment of a baseline structural condition of the pipeline, a comparison of this to the initial design condition, and a forecast of future condition, based on observed corrosion and environmental factors. In this type of analysis, structural evaluation and assessment of factors of safety are key to making decisions of remaining service life.

Other qualitative approaches have been used based on observed condition and the use of deterioration curves, however, the structural condition of the pipeline should also be considered when implementing the approach.

11.5 RISK RANKING ANALYSIS

Risk analysis is an approach to prioritizing and planning inspection, rehabilitation, and/or replacement of pipelines. This type of analysis considers the probability (or likelihood) of a failure and the consequence of failure of an asset to calculate its risk of failure. The level of risk is used to develop a prioritized plan for action and is typically used for capital improvement planning in an asset management program. The purpose of the condition evaluation in the assessment program is to determine the likelihood of failure. The consequence of failure would likely be determined during the planning stage (see "Planning" chapter).

11.5.1 Condition Ranking

The condition ranking considers the probability or likelihood of a failure. As water pipelines age, their condition deteriorates. This deterioration is identified and quantified through the use of the various condition assessment tools identified in previous chapters. Interpretation of the observed condition and translation into a condition rating is an accepted procedure. A number of methodologies to assign numerical values to the condition of a pipeline have been developed and are widely used. Efforts are being undertaken to develop and standardize a universal ranking system, but these are still in the formative stages at the time of this writing.

Condition Description	Ranking
Very Good (0% maintenance required)	1
Minor Defects (5% maintenance required)	2
Moderate Maintenance Required (10-20% maintenance required)	3
Requires Renewal (20-40% maintenance required)	4
Unserviceable (>50% maintenance required)	5

Fig. 11-4. Example of condition ranking

Environmental factors should be considered in the condition ranking, such as geotechnical consideration, soil corrosivity, water aggressiveness, and loading conditions, as well as the results of structural and hydraulic analyses. An example of condition rankings is shown in Fig. 11-4.

11.5.2 Criticality Ranking

The criticality ranking is intended to capture the fact that consequences of failures vary throughout the system. Depending on the location of a pipeline within the system, the consequence of a failure will be much different. A failure of a major transmission pipeline, which conveys water through busy downtown areas at a very high pressure, could affect thousands of customers, cost millions of dollars to repair, and possibly millions of dollars in claims due to collateral damage. A failure of that same transmission pipeline in a rural area would have a much lower consequence of failure, and thus its criticality ranking would be much lower. An example of condition ranking is shown in Fig. 11-5.

11.5.3 Risk Ranking

Risk is computed by multiplying a numerical estimation of the condition ranking by criticality ranking. This results in a relative ranking of the pipe segments. The segments with the highest overall risk ranking are prioritized for action first. This type of analysis is useful for both improving system reliability and minimizing the adverse effects of collateral damage. This type of analysis results in the worst pipe, which can cause the most

Level of Service / Criticality Category	Weight	Negligible = 1	Low = 4	Moderate = 7	Severe = 10
Health & Safety of Public and Employees	15%	No injuries or adverse health effects	No lost-time injuries or medical attention	Lost-time injury or medical attention	Loss of life or widespread outbreak of illnesses
Compliance with Regulations	25%	100% compliance with permits	Violation but no enforcement action	Violation with minor enforcement action	Enforcement action with fines
Financial Impact on Utility	15%	Absorbed within budget line item < $5,000	Absorbed within current budget $5,000 to $25,000	May require transfer from reserves $25,000 to $150,000	May require new borrowing or impact rates > $150,000
Public Confidence	20%	No events or loss of public confidence	Minor event with limited news coverage	Short-term event; substantial loss of public confidence	Long-term event; major news event permanent loss of public confidence
Service Delivery	25%	No breaks, outages, or T&O complaints	No breaks or outages, infrequent T&O complaints	Short duration breaks or outages; occasional T&O complaints	Numerous breaks, outages; widespread T&O complaints

Fig. 11-5. Example of a criticality matrix
Note: T & O = taste and odor.

Failure Probability	Consequence				
	Insignificant	Minor	Moderate	Major	Severe
Almost Certain	MODERATE	HIGH	HIGH	EXTREME	EXTREME
Likely	MODERATE	MODERATE	HIGH	HIGH	EXTREME
Possible	LOW	MODERATE	MODERATE	HIGH	HIGH
Unlikely	LOW	LOW	MODERATE	MODERATE	HIGH
Rare	LOW	LOW	LOW	MODERATE	MODERATE

Fig. 11-6. Example of a risk ranking matrix

damage, being given the highest priority. A sample ranking matrix is shown in Fig. 11-6.

11.6 USING THE CONDITION ASSESSMENT REPORT

The condition assessment report is an important tool in the development of an asset management program. It provides the basis for addressing alternatives analysis, economic assessment, and prioritization of improvements. Although these are not elements of a condition assessment report, it is important that the report provide sufficient information so that these functions can be intelligently addressed in the program.

11.6.1 Alternatives Analysis

An evaluation of alternatives (operational changes, preventative maintenance, rehabilitation, or replacement) should be performed to determine the most feasible course of action based on the results of the analyses performed on the pipeline. The lines among preventative maintenance, rehabilitation, and replacement can be blurred. Preventative maintenance can include aspects of repair to the pipeline, but when repair becomes extensive, evaluating full rehabilitation or replacement is necessary.

With operational changes, the condition under which the pipeline is operated is changed to reduce the risk of a failure, e.g., a reduction in pressure. With repair, pipeline functionality and reliability are restored. Rehabilitation aims at restoring the pipeline close to its original operating condition, whereas pipeline replacement completely replaces the old pipeline with new. Some examples of alternatives for each category that can be considered are as follows:

- Operational Changes
  - Lower operating hydraulic gradeline
  - Reduce anticipated surge pressures
  - Adjust operations to increase redundancy
- Preventative Maintenance
  - Install internal joint seals
  - Install cathodic protection
  - Schedule frequent inspections
  - Replace key components (valves)
  - Identify causes of surge or transient pressure and design them out
- Rehabilitation Technologies
  - Cleaning and lining
  - Curing in place pipe
  - Slip lining
  - Carbon fiber lining
  - Pipe bursting
  - Tendon repairing
- Replacement
  - Trenchless
  - In place
  - New location
  - In situ

11.6.2 Economic Assessment

Cost is often the driving factor when choosing a solution to mitigate a pipeline at the end of its designed service life. Once solutions are considered, an economic assessment should be conducted. In cases in which multiple feasible solutions are identified, a comparative life cycle analysis should be conducted to select the best solution from an economic standpoint that still meets other critical objectives, e.g., reliability, redundancy, and resiliency. Ideally, the inspection results and alternative analyses will be used to select the most cost-effective ways to provide the desired level of service. A life cycle cost analysis should be used to compare two alternative strategies (e.g., replacing a pipe or rehabilitating it). A costs-benefit analysis (economic, environmental, and social) can be used in some cases and incorporated into the risk analysis.

The alternatives that are economically analyzed should be consistent with meeting the utility's risk tolerance. A cost analysis may show that preventative maintenance of old pipes and periodic point repair in areas of low failure consequence are less expensive than building new pipes.

11.6.3 Prioritization of Improvements

The final outcome recommended should be a consolidated resource plan to be implemented by the utility to address the needs identified in the condition assessment. Once all pipe segments are ranked, solutions identified, and costs estimated, the utility can incorporate the results into its capital outlay program. High priority pipelines may need to be worked into the utility's short-term capital improvement budget, whereas those ranking lower may be incorporated through long-term improvements planning.

CHAPTER 12
CASE STUDIES

The following case studies are presented to provide examples of various condition assessment projects using some of the technologies previously discussed. As can be seen, multiple technologies are often used for a project to optimize the testing results, depending upon pipe material and size, expected pipe condition, whether the pipe was in or could be taken out of service, and other factors.

Presented are seven projects incorporating a variety of condition assessment technologies. In each case, the results obtained were found to be successful and reasonably conclusive in determining each pipeline condition and useful in determining pipeline rehabilitation and/or replacement recommendations.

The case studies included are

- Weber Basin Water Conservancy District in Utah,
- Miami-Dade Water and Sewer Department in Florida,
- Tarrant Regional Water District in Texas,
- San Diego County Water Authority in California,
- Pinellas County Utilities in Florida,
- Dallas Water Utilities in Texas, and
- City of Houston Department of Public Works and Engineering in Texas.

12.1 WEBER BASIN WATER CONSERVANCY DISTRICT, UTAH: WATER SUPPLY AQUEDUCTS CONDITION ASSESSMENT USING MULTIPLE INSPECTION TECHNOLOGIES

Project Owner Weber Basin Water Conservancy District	**Technology** Smart Ball, Pipe Penetrating Radar, Sonar, Ultrasonic, Panoramo, and Conventional CCTV
Pipeline Material(s) Reinforced Concrete, Cement Mortar Lined and Coated Steel, and Bar Wrapped Concrete Cylinder Pipe	**Project Date** 08/2010 to 06/2012
Consultant Carollo Engineers Halcrow (Risk Model) Infinity Corrosion	**Project Cost** $801,500
Contractor N/A	**Owner Contact** Adam McKnight, P.E. Project Manager 2837 East Highway 193 Layton, Utah 84040 amcknight@weberbasin.com

12.1.1 Project Overview and Justification

The Weber Basin Water Conservancy District (WBWCD) provides treated municipal water to more than 600,000 residents as well as wholesale irrigation water, retail secondary irrigation water, untreated industrial water, and groundwater replacement water for the Northern Utah region. The Weber and Davis raw water aqueducts can supply an estimated 250 million gallons of raw water per day to three water treatment plants, agriculture, and industrial users. As the solitary water supply to the community, planned shutdowns are limited to four days in the fall during low flow conditions.

WBWCD's 26 mi of pipelines were constructed in 1955, composed of reinforced concrete (RCP) pipe, cement coated and lined steel pipe, and bar wrapped concrete cylinder pipe. In the past, the aqueducts have exhibited leaking at some of the joints along with one past catastrophic failure in the RCP. The failure was due to a debris flow that moved a segment of the Davis Aqueduct. The pipeline ranges from 84-in. to 21-in. diameter as it approaches North Salt Lake City. Fig. 12-1 is a vicinity map of the two aqueducts.

WBWCD employed Carollo Engineers to perform the first condition assessment of the aqueducts, perform a risk analysis, prepare a hydraulic model, create a projected maintenance schedule, and recommend a feasible five-year capital improvement program (CIP).

Until this project, the aqueduct system was continually in service. Shutdowns were limited to the time necessary to repair the aqueduct. This project successfully orchestrated a four-day shutdown to perform targeted internal inspections on 5 mi of the aqueducts. The project involved

Fig. 12-1. Weber and Davis aqueducts vicinity map
Source: Courtesy of Carollo Engineers

five internal (and one external) technologies for inspection testing: nontethered leak detection and entrapped pockets of air detection by SmartBall, without a shutdown; high-speed digital three-dimensional (3D) video scanning by Panoramo; conventional closed-circuit television (CCTV); sonar; in pipe penetrating radar (PPR); and ultrasonic testing (UT) readings (exterior). Based on the risk model developed for the system, approximately 5 mi of the 26 mi was inspected during the shutdown.

External investigations included UT on steel pipe, hammer sounding, geotechnical investigation (entire alignment), electrical conductivity, and stray currents.

12.1.2 Technologies

Nontethered Leak Detection and Entrapped Pockets of Air Detection with SmartBall The SmartBall is an acoustic-based technology that detects anomalous acoustic activity associated with leaks or air pockets of trapped gas in pressurized pipes. The SmartBall is composed of a watertight, aluminum alloy core that contains a power source, electronic components, and instrumentation (including an acoustic sensor, accelerometer, magnetometer, GPS synchronized ultrasonic transmitter, and temperature sensor). The aluminum core is encapsulated inside a protective outer foam shell that provides additional surface area to propel the device through the pipeline as well as reduce some low frequency ambient noise that is typically present in the pipeline. The SmartBall assembly is deployed through a 4-in. diameter entry into the flow of a pipeline to traverse the pipeline and is captured at a point downstream.

Fig. 12-2. SmartBall insertion near the bifurcation structure
Source: Courtesy of Carollo Engineers

Weber Aqueduct On August 3, 2010, coincidental with high water demands, a SmartBall was inserted into the 48-in. aqueduct near the bifurcation structure through a 4-in. gate vale installed on a 20-in. access man way and retrieved at an air release valve (ARV) located 7,450 ft away (Fig. 12-2). The internal pressure along the run varied from 8 psi at the insertion point to as much as 200 psi in the low area of the large siphon. Acoustical data collected by the SmartBall as it traversed the aqueduct was used to evaluate and identify acoustic anomalies associated with leaks and air pockets.

The survey revealed one anomaly resembling a leak, and one as a pocket of trapped air estimated at 21 ft in length. The identified leak was a simulated leak created by WBWCD staff at a blowoff valve without knowledge of the field personnel.

Davis Aqueduct On August 4, 2010, coincidental with high water irrigation demands, a SmartBall was inserted into the 60-in. portion of the aqueduct at a 4-in. gate valve in a 20-in. man way, with plans for retrieval approximately 24,860 ft downstream at the terminal structure. The SmartBall was finally retrieved on November 15, 2010, after being trapped in deposition after running approximately 16,900 ft in the aqueduct (Fig. 12-3). The battery was dead, but the data still remained in the instrumentation of the unit.

Three acoustic anomalies that resembled leaks were identified. Two of the leaks were simulated by the WBWCD to test the accuracy of the technology. Both were classified as medium-size leaks. The latter leak was detected, but the location was not confirmed. The third leak was also classified as a medium-size leak and was determined to be a valve that was not properly seated. The leak detection testing also identified significant deposits of debris in a section of the aqueduct.

Panoramo CCTV Compared with traditional CCTV, the Panoramo digital pipeline scanners can inspect twice as much pipe per day, increase the image resolution six times, and reduce operating costs by as much as

Fig. 12-3. Deposits in 48-in. aqueduct at North Canyon inverted siphon
Source: Courtesy of Carollo Engineers

30%. At the core of the system are the wide angle (185°) twin digital cameras. The digital cameras are installed on the front and rear of the scanner to capture still images as the system moves rapidly through the pipeline. Extremely powerful xenon strobes illuminate up to 48-in. pipelines, and the still pictures captured by the cameras produce high-resolution digital images that do not suffer from motion blur at high inspection speeds. The Panoramo system captures a 3D image of the pipeline and completely eliminates the pan and tilt functions of a traditional camera. This creates a distinct advantage for the Panoramo camera, by increasing the speed of the inspection and capturing a complete image of the pipeline. In addition, the pipeline image rendering is done in the office, eliminating missed defects and reducing the impact to the public.

Weber Aqueduct The inspected segment consisted of a 48-in. concrete cylinder pipe with medium pressure. System pressures varied from 160 to 190 psi. Very little sediment was found in the Weber Aqueduct; however, 3/8 in. of slime was observed on the pipe wall. The slime consisted of fine sand, silt, and clay that had adhered to the pipe wall. Approximately 2,930 ft of the 48-in. pipe wall was pressure washed, and the condition of the pipe was examined using Panoramo CCTV for documentation. Access was gained through 20 in. man ways at the top of the pipe. No problems were noted with this portion of the pipeline.

Davis Aqueduct This segment consisted of 33-in. and 21-in. concrete cylinder pipe with medium pressure. System pressures varied from 0 to

Fig. 12-4. Longitudinal and circumferential cracks in crown of pipe
Source: Courtesy of Carollo Engineers

165 psi in this segment. Inspections in this segment were made primarily with a Panoramo camera. A total of 1,600 ft of camera inspection occurred in this segment. Longitudinal and circumferential cracking at the crown of the pipe (shown in Fig. 12-4) was observed in a section of the pipeline.

Further investigations are planned to determine the extent of the damage and possible rehabilitation strategies.

Sonar Sonar is used to identify problems below the flow line of a pipe and is used extensively in siphons that cannot be dewatered. Sonar units emit a sound wave that travels until it hits a solid object (pipe wall or debris line) and returns back to the unit. The sound wave outlines the shape of whatever the wave bounces off. The information can be used to identify the level of debris or whether the pipe could be losing structural integrity by observation of the *roundness* of the pipe. In addition, the quantity of debris can be calculated using this technology.

Davis Aqueduct The initial concept was to inspect approximately 2,500 ft of the aqueduct with sonar technology, but during the shutdown and inspections, the authors were ahead of schedule and were able to dewater sections of the pipe for manned entry and debris cleaning. The total sonar footage was reduced to 830 ft of 84-in. diameter pipe. The sonar investigation of the three siphons revealed the pipeline to be clean, which is surprising considering that the siphons had never been cleaned.

Pipe Penetrating Radar (PPR) PPR is the underground in-pipe application of ground penetrating radar, a nondestructive testing method that can detect defects and cavities within and outside mainline diameter

[>18 in. (450 mm)], nonmetallic [i.e., RC, vitrified clay, PVC, high-density polyethylene (HDPE), etc.] underground pipes. The key advantage of PPR is the unique ability to map pipe wall thickness and deterioration including voids outside the pipe, enabling accurate predictability of needed intervention or the timing of rehabilitation and/or replacement.

Davis Aqueduct The testing was done manually, using an inventory grid with 2-in. spacing within the pipe. The initial testing was performed on four joints in the 60-in. RCP (Fig. 12-5). Three locations were known leaking joints that were repaired by mechanical seals. To conduct the inspection on the leaking joints, the mechanical seals were removed. PPR tests were also conducted on a control joint that was not leaking for comparison. The resulting imagery produced with the PPR technology requires trained specialists with specialized software to interpret the results.

The PPR revealed ambiguities in the three known leaking joints that appear to be joint cracks and two of the joints had voids consistent with past leaks. The sound or control joint results showed no ambiguities, which validated the results of the survey.

Ultrasonic Testing UT is probably the most established material testing technique for assessing ferrous pipe wall thickness and conditions. The technique involves the travel of ultrasound through the metal pipe from the external surface inward. The pulse of sound will progress through the metal pipe at a certain velocity until a boundary surface causes partial or total reflection. The return pulse is collected by the receiving transducer. The distance is calculated by the unit software and

Fig. 12-5. PPR inventory grid on 60-in. RCP
Source: Courtesy of Carollo Engineers

displayed digitally. Measurements are typically taken on a grid system to enable the inspection to be carried out in a systematic way. It is to be noted that flaws or inclusions can produce echoes, which may cause erroneous wall thickness measurements.

Weber Aqueduct One excavation on the 48 in. cement lined and coated steel pipe was made for the investigation. UT readings taken on the outside of the pipe indicated that the steel pipe was in excellent condition. No corrosion was observed.

Davis Aqueduct One large excavation, which uncovered more than 20 ft of 33-in. bar wrapped concrete cylinder pipe (C303), was made for the investigation. UT readings taken on the outside of the pipe indicated that the steel cylinder was in excellent condition. The exterior mortar coating was still causing passivation of corrosion on the steel cylinder. No corrosion was observed.

12.1.3 Recommendations

As part of the investigation, the team observed 5% of the total pipeline joints using various technologies and man entry observations. Upon completion of their investigation, it was determined that 23% of the joints observed need repair or rehabilitation. If the joints inspected were representative of the entire aqueduct, approximately 1,900 joints will need to be rehabilitated to help ensure that the joints maintain the estimated remaining life of the reinforced concrete and steel pipeline barrel of 100 years.

The inspection team assessed both the viability of each of the inspection technologies for future monitoring of pipeline conditions as well as the conditions of the aqueduct itself.

Based on the field observations, technical evaluations and deterioration analysis, hydraulic modeling, and risk assessment, the project team prepared a five-year CIP. The CIP addresses

- Pipe joints (RCP) and potential pipe barrel rehabilitation in the 21 to 33-in. concrete cylinder segments;
- Isolation valves, upsizing existing ARVs, and air vents; and
- Budgetary capital improvement and future condition assessment costs.

In summary, the project was of great benefit to the WBWCD, which continues a proactive inspection-renewal cycle of its infrastructure. The condition data are now being used to formulate rehabilitation strategies. The experience gained on this project will help ensure cost-effective selection of inspection technologies in the future.

12.2 MIAMI-DADE WATER AND SEWER DEPARTMENT, FLORIDA: PIPEDIVER INSPECTION

Project Owner Miami-Dade Water and Sewer Department	**Project Date** May 2010, ongoing
Pipeline Material(s) Prestressed Concrete Cylinder Pipe	**Project Cost** Ongoing
Consultant Pure Technologies	**Owner Contact**
Contractor N/A **Technology** PipeDiver	 Source: Courtesy of Pure Technologies U.S.

The Miami-Dade Water and Sewer Department (WASD) conducted an electromagnetic (EM) condition assessment using PipeDiver. The inspection consisted of 8 mi of 48-in. and 54-in. prestressed concrete cylinder (PCC) pipe built in the mid-1970s. The pipeline is used for treated water transmission and serves more than one-third of the county with treated water. Miami-Dade WASD is one of the 10 largest water utilities in the United States, serving approximately 2.5 million people and pumping more than 300 million gallons of water per day. The Miami-Dade inspection, in May 2010, was the first commercial inspection using this technology in the United States.

In March 2010, Miami-Dade had two failures: one on a 54-in. PCC water transmission main, causing a 40 × 40-ft sink hole in the middle of an intersection (Fig. 12-6); the other on a 72-in. PCCP sewer force main. These failures prompted Miami-Dade WASD to find and use the best possible pipeline inspection technologies to assess and address the problems in their water and wastewater pipelines.

For the EM inspection, WASD used the PipeDiver tool, which allowed for nondestructive EM assessment of the transmission main. This tool is free swimming and composed of three parts: a battery module, an EM module, and a tracking module. It has EM sensors on each fin that take readings as it flows throughout the pipe with the water. This inspection was completed in two runs, with four total access points. The access points were 18-in. boiler-type man ways that required excavation to access.

The inspection was primarily completed at night because the pipeline crossed a number of heavily trafficked highways and roads and was inaccessible during the day. The tool was tracked at ground level with

Fig. 12-6. 54 in. PCCP failure
Source: Courtesy of Pure Technogies U.S.

a handheld sensor—making safety a primary concern and reason for inspections to be conducted at night.

This pipeline presented several locations of concern, where contingency measures were needed to ensure a successful run. At the locations where the pipeline deviated from the highway to urban areas, it included several tees used for 90° bends with a blind flange at the other end in the case of future expansion. These consecutive sharp 90° turns created the risk of the tool stopping and getting lodged in a bend. To mitigate this, the inspection team (including a team of commercial divers) was ready at the locations, in case the pipeline needed to be depressurized and the tool redirected to the flow path manually. The contingency measures were not needed; the tool was able to continue its travel along the flow path with no problems traveling through the consecutive bends.

Finally, the pipeline structure did not support the tool insertion and extraction tubes—it was not a typical manhole with easy ground access. To accommodate, the inspection team excavated pipe sections for insertion and extraction, depressurized the pipeline, and inserted the tool. For extraction, commercial divers entered a short segment of the pipeline to retrieve the tool when it arrived at the extraction location.

The EM data collected was analyzed by comparing the EM readings taken inside the pipeline to a calibration curve. The calibration curve shows the EM levels of different amounts of wire breaks in a pipe section based on previous PCCP projects or from test pipes. For this inspection, calibration curves were created based on a pipe section that had been previously removed. The data analysts could then make a wire break estimate for each pipe section, based on the EM reading compared against the calibration curves.

Of the 4,504 pipes that were inspected, 126 showed distress. The majority of the distressed pipes were found along the 54-in. portion of the transmission main—with a distress rate of 9.2%, compared with only 0.2% on the 48-in. portion. It was found that only 1.6% of the entire pipeline was at high risk, which was consistent with other inspections across North America. However, 5.5% of the pipe sections along the 54-in. portion were at high risk, which is notably higher than the average.

It was recommended that the distressed pipe sections be repaired or replaced in priority based on the level of risk, sensitive to budgetary restrictions. In addition, the operating pressure was analyzed and recommendations were provided to reduce the risk of failure from exceeding the maximum pressure level allowable. For priority areas that could not be addressed immediately, acoustic fiber optic (AFO) monitoring was recommended as an option to track wire breaks and reduce the risk of failure.

The inaugural commercial free-swimming inspection in Miami-Dade led to the development of a comprehensive pipeline management program for Miami-Dade WASD, which continues to perform assessments of water transmission and sewer force mains to reduce its risk of failure. In addition, Miami-Dade WASD has installed an AFO system in its large-diameter pipelines (beginning in late 2012) to monitor pipeline deterioration and proactively address at-risk pipes. Over an 18-month period, Miami-Dade WASD assessed more than 100 mi of water transmission and wastewater force mains. The WASD found that only 0.7% of the pipes inspected required repair, whereas more than 99% can be safely managed.

12.3 TARRANT REGIONAL WATER DISTRICT, TEXAS: CONDITION-BASED MAINTENANCE OF A WATER TRANSMISSION SYSTEM

Project Owner Tarrant Regional Water District	**Consultant** Pure Technologies Simpson, Gumpertz & Heger
Pipeline Material(s) Prestressed Concrete Cylinder Pipe	**Project Cost** $
Technology Smart Ball, Pipe Penetrating Radar, Sonar, Ultrasonic, Panoramo, and Conventional CCTV	**Owner Contact** David H. Marshall, P.E., Courtney Jalbert, Jason Gehrig, P.E.
Project Date 1993 to Date	

The Tarrant Regional Water District (TRWD) is a raw water supplier in North Central Texas, serving communities in 11 counties. TRWD has 29 municipal customers, with Fort Worth being its largest customer. The municipal customers deliver treated water to approximately 1.8 million people. The water use in the service area is increasing 1.4% annually due to population growth. Water use in Tarrant County is approximately 92% of the TRWD's total use. The TRWD's supply, shown in the map (Fig. 12-7), consists of four supply reservoirs, two pipelines from the eastern reservoirs, a pipeline connecting Eagle Mountain Lake, and two additional pipeline terminal storage reservoirs. The reservoirs produce approximately 511,400 acre-ft of firm supply. TRWD's next supply is a wetlands water reuse project at Richland-Chambers that will add 63,000 acre-ft to the system. The two western reservoirs, Bridgeport and Eagle Mountain, hold approximately 20% of the total supply and release water down the Trinity River to Tarrant County. Water from Cedar Creek is pumped (using a 72-in. diameter, 68-mi long pipeline) and gravity flows from a balancing reservoir (using an 84-in. diameter, 6-mi long pipeline) to Tarrant County. Water is pumped from Richland-Chambers in a 90-in. diameter, 72-mi long

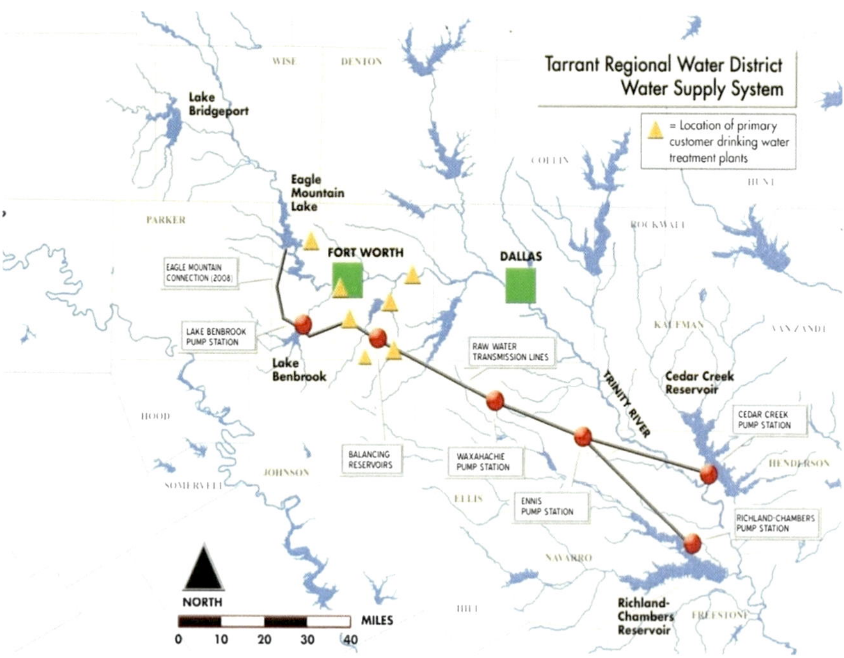

Fig. 12-7. TRWD water supply system overview map
Source: Courtesy of Pure Technologies U.S.

pipeline and gravity flows (using a 108-in. diameter, 6-mi long pipeline) to Tarrant County. The Arlington, Eagle Mountain, and Benbrook lakes serve as terminal storage reservoirs for the pipeline. Water from Lake Benbrook can be pumped into the pipeline system to serve the Tarrant County water treatment plants. Because Cedar Creek and Richland-Chambers reservoirs are 80% of the water supply, the reliability of the pipelines is essential.

12.3.1 Pipeline Integrity Issues

The Cedar Creek Pipeline was constructed in 1972 using PCCP. The pipeline has suffered 14 failures, the first in 1981. Nine of the failures have been due to corrosion of the prestressing wires. Four failures have been due to hydrogen embrittlement damage. One failure was a shear failure on a segment adjacent to a section bedded in concrete. All have been catastrophic failures, resulting in the loss of millions of gallons of water. One area of the Cedar Creek Pipeline had an impressed current cathodic protection (CP) system, and the 1.2-volt driving force damaged approximately 30% of the surrounding pipes due to the wire embrittling. The Richland-Chambers Pipeline was constructed in 1988 using PCCP. The Richland-Chambers Pipeline has suffered 15 failures: four thrust restraint failures, seven corrosion failures, two due to hydrogen embrittlement, one from a spigot cracking, and one from a tractor operator error. All but the thrust restraint and spigot cracking were catastrophic failures. Richland-Chambers had no CP, and the embrittled segments were probably because of the wire characteristics being influenced by light surface corrosion of the wire under the haunches, where mortar failed to flow. The Benbrook Pipeline is also PCCP; it was constructed in 1998 and has no integrity issues.

12.3.2 Causes of the Failures and Mitigation

Almost all the failures of the pipe segments were a result of corrosion or embrittlement. The insufficient thrust restraint on the Richland-Chambers Pipeline required redesign of the restrained pipes using thicker cylinders in the replacement pipe. Some of the Cedar Creek pipes were constructed with wire that was strain aged and had very high tensile strengths; a few tensile strength tests exceeded 310,000 psi. The wire was determined to be very susceptible to hydrogen embrittlement through testing. The Richland-Chambers Pipeline was manufactured during the time that the specification for prestressing wire fabrication, ASTM A648, was modified. The initially manufactured pipe has some susceptibility to embrittlement because it was manufactured under the old specification, but the majority of the line has wire that is less susceptible thanks to the specification change.

Both pipelines have experienced very high pressures due to transient waves. Transient control is handled by backflow through the pumps with controlled closing of the pump control valves. Many times the hydraulic actuators of the pump control valves have malfunctioned, closing the valves prematurely. Because of the high pressure, the outer mortar coating of both pipelines has been cracked, allowing a pathway for water and corrosion to take place. Soils in the area have very low resistivity, less than 2,000 ohm-cm. The active soil, coupled with cracked coating, provided for rapid degradation of the prestressing wire.

To mitigate the failures, three paths were taken: pipe segment replacement, CP, and transient control. The pump control valves were modified so a programmable logic control timed the valve closing, reducing the transient wave dramatically. CP was accomplished using zinc anodes attached to the pipelines, developing a target potential of −0.85 volts. To retrofit both lines took approximately 5 years, during which time 33,000 zinc anodes (60 lb each) were installed.

Segment replacement was initially based on visual inspection. If longitudinal cracks were found, the pipe was replaced. TRWD had the opportunity to visit the San Diego County Water Authority (SDCWA) in 1992 and witness the sounding inspection techniques being used there. Using soundings in Texas, TRWD was able to find damaged segments before any cracking became apparent. With the advent of EM inspection, the number of broken wires was able to be quantified. TRWD joined with seven other agencies to employ Simpson, Gumpertz & Heger (SGH) to develop risk curves and better define replacement based on wire breaks due to corrosion. TRWD continued this work, also evaluating embrittled pipe damage. To date, TRWD has replaced 121 segments on the Cedar Creek Pipeline and 115 segments on the Richland-Chambers Pipeline.

Inspections When the frequency of pipe failures began to increase in the late 1980s, the inspections were conducted in areas that were dewatered during the repair of a failure. The pipe was examined for cracks. Starting in 1993, sections of the pipeline were deliberately dewatered and inspected using soundings. In 1996, TRWD took part in a study funded by the American Water Works Research Foundation (now the Water Research Foundation) to test nondestructive inspection techniques for PCCP including visual, soundings acoustic monitoring, and acoustic reflection. TRWD used the acoustic monitoring technique and did find distressed pipe. In 1998, TRWD employed Pressure Pipe Inspection Company (PPIC) to inspect, using their remote field eddy current (RFEC) coupled transformer inspection technique, and found the technique, when calibrated, to provide a quantified count of broken wire.

TRWD has inspected all 164 mi of prestressed pipe, taking 10 years to do so. A total of 1,283 damaged pipe segments were found, approximately 2.7% of the 47,666 segments in the system. There are currently 1,047 segments in the system with damage. Most are embrittlement damage. The damaged pipes are reevaluated annually, some with reinspection and all with an estimate of continued deterioration and prioritized for replacement. The graphs (Fig. 12-8) show the spatial distribution of the damaged segments.

12.3.3 Integration of Geographic Information System and Pure Technologies' Inspection Results

TRWD employed Pure Technologies, formerly PPIC, to map out the existing pipeline system in the early 2000s to provide an accurate map of the pipeline. TRWD's as-builts and lay sheets proved unreliable for pipe details and position of bends. They also proved problematic during replacement because there was always uncertainty and not always visible signs of distress. As a result, inspections were done for TRWD's entire system over time and used as a guide to create a geographic information system (GIS) model by counting pipe and comparing with record drawings. As a result, TRWD now has an accurate map of the pipeline segments, appurtenances, and inspection results. The GIS model also makes it possible to verify pipes for pipe replacement and is used to make decisions for possible future replacements.

At that time, tables were also created for storage of not only pertinent pipe information but of all the inspection information collected. With each inspection, there are six tables that are updated, reviewed, and added to TRWD's GIS model. Using these tables, TRWD is able to track damage for each individual segment and compare trends in damage from one inspection to another.

After each inspection, each pipe with broken prestressing wire is updated within TRWD's database and a new record is added to the inspection history. The model keeps track of each inspection and is assigned an individual inspection ID. The data stored for each inspection also includes dates of the inspection, the inspection staff members present, comments, total wire breaks, the linear X ft along the pipe where damage is found, and the number of wire breaks for each linear X ft. Because of improvements in technology over the years, Pure Technologies also reanalyzes any past inspections in the same area and adjusts accordingly.

Since the model began, TRWD has also added pipe-to-soil potentials, soil resistivity, replaced pipe segment information, test station readings, and anode information.

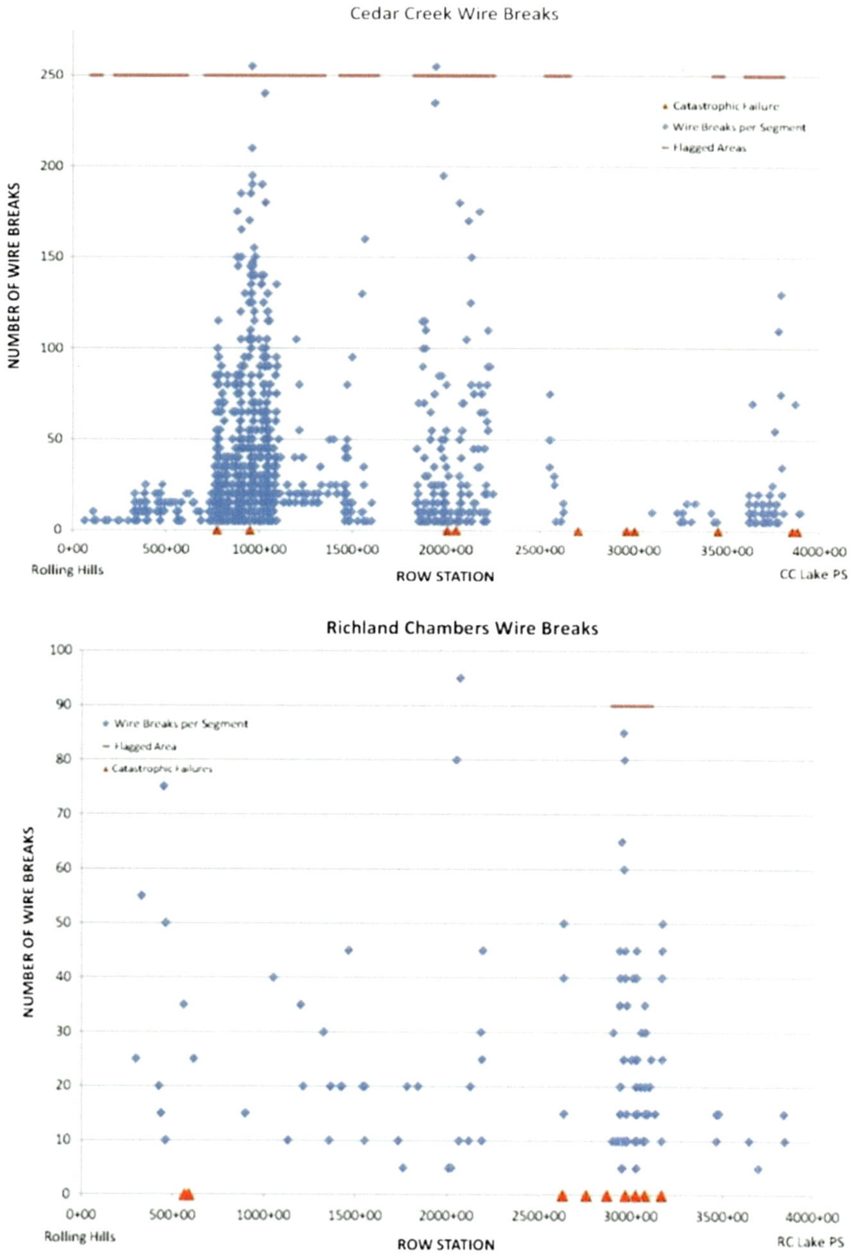

Fig. 12-8. Spatial distribution of distressed pipe segments for Cedar Creek and Richland-Chambers Pipelines
Source: Courtesy of Pure Technologies U.S.

12.3.4 Integration of SGH and Pure Technologies' GIS Prioritization Tool

TRWD was tasked with prioritizing the damaged pipes and began to change the replacement program from one that replaced highly distressed segments to one of managing risk by, instead, replacing segments that are the highest risk of creating problems, using a prioritization scale. This is when the task of creating a GIS tool that would incorporate the inspection data, currently in TRWD's database, with a simplified strength model was assigned. The tool was developed to help determine which segments needed to be replaced and when. It uses a simplified strength model to examine the impact of adjacent corroded and failed prestressing wire on the remaining strength of the pipe.

The tool is based on the SGH failure risk analysis and prioritization scale. First, an uncertainty analysis is performed to evaluate the effective number of broken wires to be used in the determination of repair priorities, taking into account different sources of uncertainties, such as inaccuracies in the EM testing, degradation of wires adjacent to the broken wire zone (BWZ) in corroding pipe, growth of BWZ with time, and interaction of multiple BWZs. Second, a risk analysis is performed accounting for the residual prestress found in pipe with broken wires due to hydrogen embrittlement. For pipes with BWZs due to corrosion, no residual prestress was considered.

For the tool to function properly a set of input tables were designed to store specific data needed for the tool to calculate a priority using the failure risk analysis. Pure Technologies' inspection table is incorporated into the SGH tool and is updated after each inspection with wire break information. Other tables include detailed pipe information such as pipe class, wire wraps, cylinder gauge, etc. Pressure scenarios with different pump operations and the envelope pressure (working plus transient pressure) are also included in a table for input.

The tool was designed to allow for several options of evaluating the pipeline. The options given are to choose which pressure scenario to evaluate, e.g., design class, maximum pressure, and whether to evaluate with or without CP; to specify a time to predict priority into the future; to analyze by one pipe or an entire line; to receive a priority or get a structural evaluation; and to decide which pipes may have hydrogen embrittlement or not.

After the tool is run, the simplified strength model then categorizes the damaged pipe into four priority levels (Fig. 12-9). Priority Level 1 is pipe in which the maximum pressure in the line exceeds the pressure that produces the ultimate strength of the outer core and yielding of the steel cylinder. This pipe should be repaired immediately or within a very short time period because catastrophic failure is possible. Priority Level 2 pipe

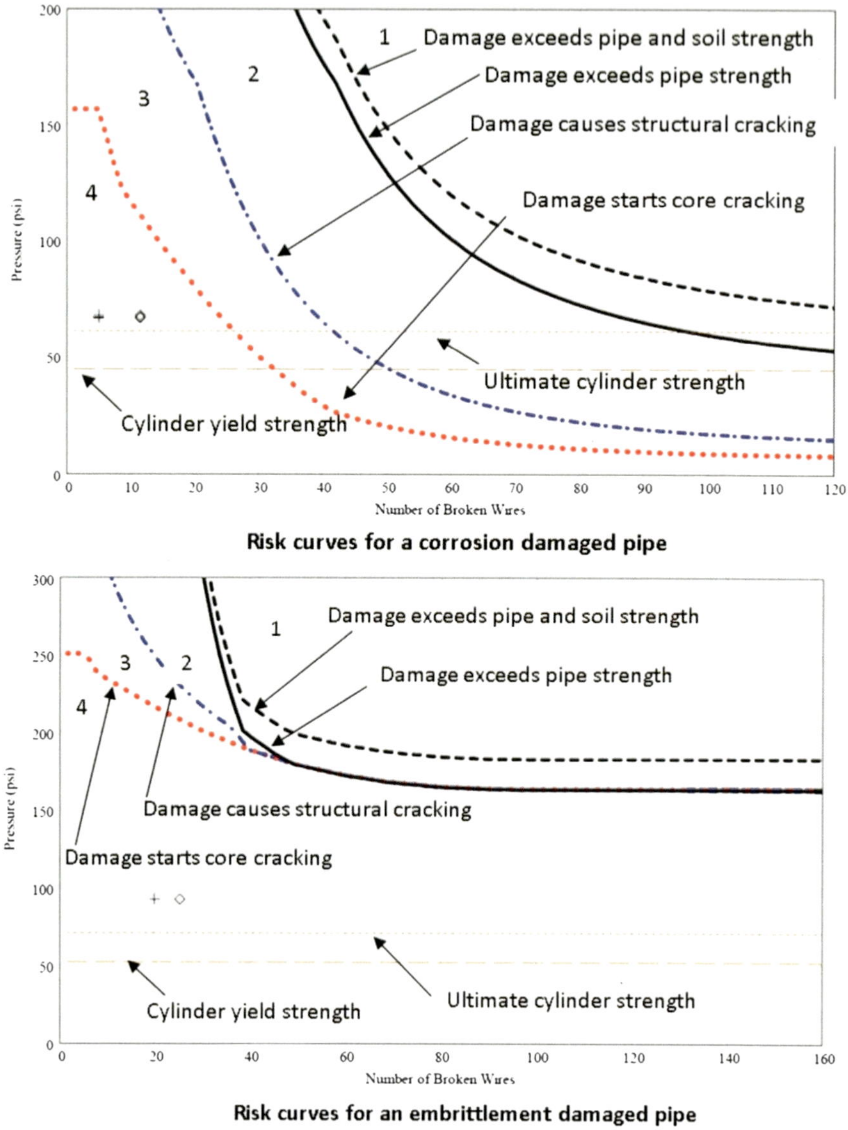

Fig. 12-9. Example of SGH risk curves for a corrosion and embrittlement damaged pipe
Source: Courtesy of Pure Technologies U.S.

is one in which the expected pressure exceeds the pressure that produces damage limit states and specifically causes a structural cracking limit state, thus exposing the steel cylinder to the environment and possible corrosion. The failure occurs with time as the number of broken wires increases, or if, and when, the steel cylinder corrodes. Repair should be performed within a reasonable time period because further deterioration of the wire or cylinder could result in failure. Priority Level 3 pipe occurs when the expected maximum pressure in the line exceeds the onset of a core cracking limit state, but not the damage limit state. The failure of the pipe, if it occurs at all, is over a much longer time period than with a Priority Level 2 pipe. A Priority Level 3 pipe should be monitored. A Priority Level 4 pipe is where the expected maximum pressure in the line is less than the onset of a cracking limit state. The failure of a Priority Level 4 pipe is not expected, and monitoring should be limited to infrequent inspection.

12.3.5 Future GIS Model Improvements

The model will be able to indicate priority and inspection information, but there are several tasks over the next few years that TRWD would like to incorporate. The first is determining what type of damage may be encountered on each segment, whether due to embrittlement or corrosion. At one point in time, some areas of TRWD's PCCP pipeline were determined to have voltage supplied (by the CP system) that was too high. Other distressed areas of the pipeline are more likely the result of corrosion due to soil type or overpressuring. Taking this into account, and using the length of damage and the number of broken wires as a guide, a decision could be made for most segments to determine which type of damage has likely occurred.

On the flip side, the model can help in determining areas of the pipeline that need CP system adjustments. Using the prioritizations and inspection data, TRWD could identify areas with too high or too low pipe-to-soil readings. Signs of corrosion are possible if potential readings from the current sacrificial anode CP system are too low and if the inspection data show a large number of wire breaks at the break location. The opposite is also true: if potential readings become too high and more extensive damage is found evenly across the pipe segment, hydrogen embrittlement is possible. TRWD would also be able to use the GIS model to track anode bed life based on rate of use of anode, number of spare anodes, and polarity.

Lastly, TRWD would like to explore tools such as the Spatial Risk Analyst by New Century Software. Software such as this would allow TRWD to create multiple risk models or implement pipeline risk models. TRWD would be able to compare risks using the SGH priorities; actual proximity of segments to facilities, homes, businesses, etc.; pressure scenarios; date since last inspection; pipe-to-soil readings; soil resistivity; areas that are difficult to repair; and many others. This tool would quantify threats and consequences along the pipeline using this information, perform the risk assessment, and then provide staff with results in the form of charts, tables, and/or risk alignment sheets. The results of such analysis would then help prioritize distressed pipe segments for replacement.

12.3.6 Results of the Program

TRWD has seen a dramatic reduction in pipe failures as a result of the mitigation done. Initial failure rates were modeled using a simple log linear relationship between the time of failure as the independent variable and the cumulative number of corrosion and embrittlement failures as the dependent variable. Based on this analysis, the Cedar Creek Pipeline could now be experiencing as many as 11 failures per year and the Richland-Chambers Pipeline as many as 30 failures per year. Fig. 12-10 shows the failure plots and log linear models.

Evaluating the failure rate using a starting date after the mitigation had been in place shows that the failure rate is reduced to 0.5 failures per year for the Cedar Creek Pipeline and 0.1 failures per year for the Richland-Chambers Pipeline. The Cedar Creek Pipeline's integrity is influenced by the quality of the prestressing wire and may never achieve the reliability of the Richland-Chambers Pipeline. Some segments of the Cedar Creek line continue to show growing embrittlement damage after the CP was applied. Evaluations of the risk of continued corrosion compared with the potential for continued hydrogen embrittlement for the Cedar Creek Pipeline showed that the CP would provide much higher reliability.

TRWD has continued to work to better control transients. On the Cedar Creek Pipeline, most of the breaks after 1996 have been associated with high pressure. The size of air valves was studied for both lines, and results indicated larger sizes were required in numerous areas. Pressure relief valves have been added in a couple locations to help reduce pressure. Connections to water treatment plants have been evaluated and control valve timing lengthened to reduce transients. TRWD will continue to explore and implement ideas to help reduce high pressure episodes.

The success of improving the reliability of the two transmission lines required spending more than $22 million over the last two decades. Current expenses are approximately $1,100,000 per year, which includes inspections, additional anode installations, and pipe replacements. These expenses will diminish some in a few years with the refurbishment of the anode system. The number of pipe replacements is expected to diminish as is the number of miles inspected annually. The costs to date represent approximately 4% of the replacement costs of the line.

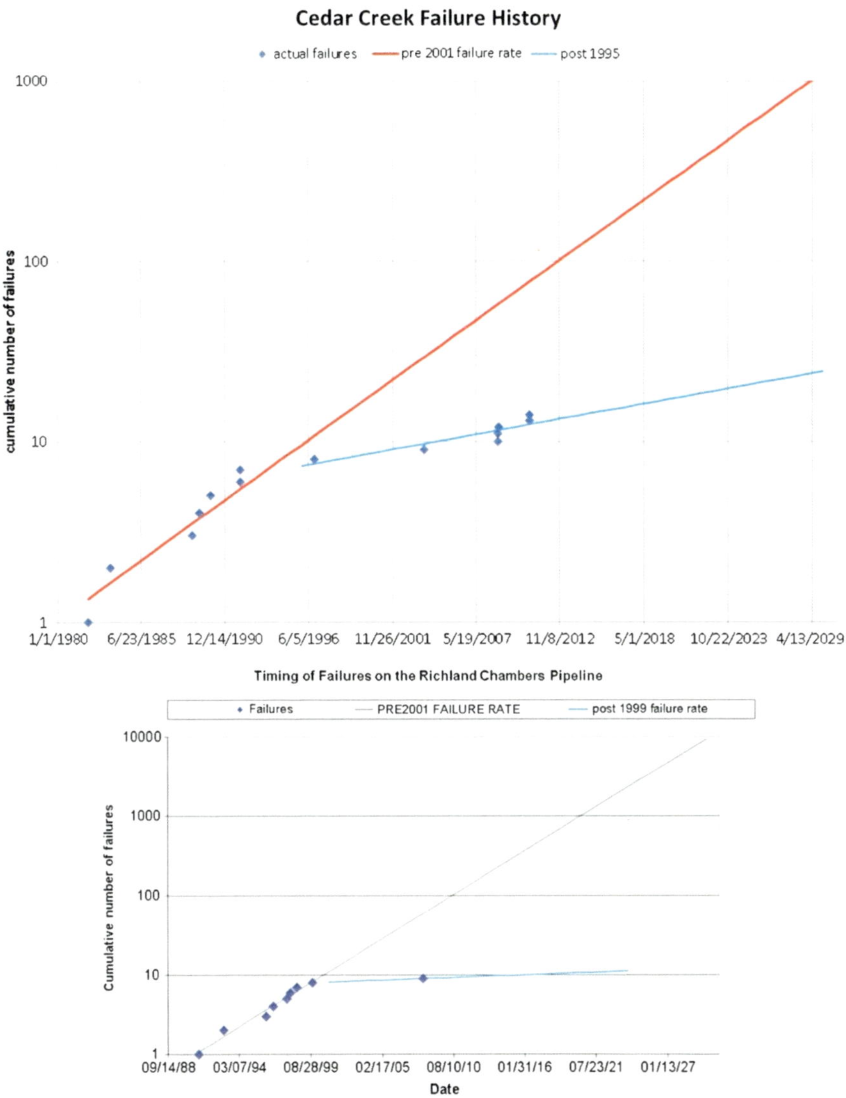

Fig. 12-10. Projected failure rates of Cedar Creek and Richland-Chambers Pipelines, before and after mitigation efforts
Source: Courtesy of Pure Technologies U.S.

Condition-based maintenance of the pipeline has proven to be extremely cost-effective and will extend the useful life of the transmission mains for decades.

12.4 SAN DIEGO COUNTY WATER AUTHORITY, CALIFORNIA: MAGNETIC FLUX LEAKAGE INSPECTION

Project Owner
San Diego County Water Authority

Pipeline Material(s)
Cement Mortar Lined and Coated Steel Pipe

Consultant
AECOM

Contractor
J.F. Shea

Technology
Magnetic Flux Leakage

Project Date
12/2011 to 03/2012

Project Cost
N/A

Owner Contact
Nathan Faber
O&M Manager
610 West 5th Avenue
Escondido, Ca 92025
nfaber@sdcwa.org

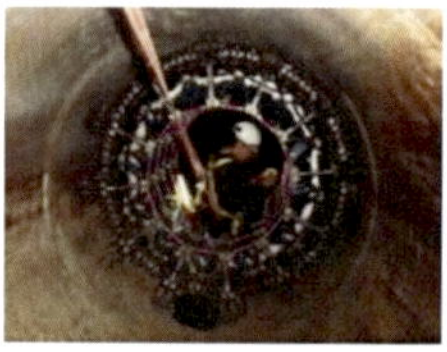

Source: Pure Technologies U.S.

Pure Technologies was employed to conduct a comprehensive magnetic flux leakage (MFL) assessment on the SDCWA's 73-in. and 76-in. mortar lined steel water transmission main, called the Second San Diego Aqueduct (SSDA). The inspection included just over 5 mi of the SSDA pipeline, which was built in the 1950s and is an important link between the Metropolitan Water District and the SDCWA.

After the Carlsbad desalination plant was approved for construction, the SDCWA needed to verify that its major transmission main could handle the additional flow created from the plant, which would eventually tie into the pipeline. A condition assessment was required to decide whether the pipeline would need to be replaced or repaired to handle the additional strain. The transmission main had been previously inspected using other techniques, like visual inspection and soil and environmental assessment, but the SDCWA needed a higher resolution assessment, as well as a quantitative analysis of the remaining pipe wall thickness, before tying the transmission main into the desalination plant.

A custom, extra high resolution MFL tool, owned by the San Francisco Public Utilities Commission (SFPUC), was used for the inspection. The MFL tool was first used by the SFPUC and Electromechanical Technologies Inc. (EMTEK), now a wholly owned subsidiary of Pure Technologies, to inspect their San Joaquin Pipelines in a project that spanned from 2007 to 2010. SFPUC loaned the MFL tool to SDCWA for the SSDA inspection.

The MFL tool is a form of nondestructive testing used to detect corrosion, pitting, and wall loss in lined metallic pipes. With MFL technology, permanent magnets are used to temporarily magnetize the steel pipe and the magnetic field changes are recorded and analyzed. The magnetic flux is uniform if there are no flaws in the wall of the pipe. If internal or external flaws are present, such as pitting, corrosion, or other forms of damage, the magnetic flux is distorted beyond the wall of the pipe, and this distortion or "leakage" is measured by Hall Effect sensors. Several different access points were used during this inspection. Both the geometry tool and the MFL tool were assembled and disassembled inside large manholes, around 16 by 14 ft in size. There were also access points every 3,000 to 5,000 ft for the safety of the inspection staff pulling the tool. Access points were also used to charge the all-terrain vehicles (ATVs) used to pull the MFL and geometry tools (Fig. 12-11), as well as to set up the winch system in areas with inclines or declines.

The condition assessment began with a geometric survey of the dewatered San Diego pipeline. The geometry tool mapped the inside of the pipeline for any distortions or structural anomalies that may have been caused by overburden or poor bedding and installation. The geometry tool measures the internal diameter of the pipeline with fingers spaced 1-in. apart. Each finger records the relative change in pipe diameter around the circumference at a rate of 500 samples per second. From these readings, an absolute distance measurement of the pipe wall diameter can be determined and represented as a deviation of pipe diameter from a normally round pipe.

After the geometric survey, the MFL tool was used to identify and locate high-resolution metal loss in the pipe wall caused by corrosion, pitting, and external pressure. The basis of a magnetic flux inspection tool is a magnetic field introduced into the pipeline wall by magnets attached to the tool. As

Fig. 12-11. MFL tool being pulled by ATV
Source: Courtesy of Pure Technologies U.S.

the tool travels through the pipeline, sensors record variations of the magnetic field around defects. MFL technology has a magnetic field so strong that the pipe wall will carry no more magnetism if the pipe wall is intact. Where there is a defect, the remaining wall is too thin to carry the full magnetic field and the flux is diverted around the defects, or "leaks" out of the pipe. During this inspection, the MFL tool was pulled slowly through the interior of the pipeline by an ATV. Because of its weight, it was pulled up inclined areas using a winch system stationed on the ground near a manhole and released down slopes using the same method. MFL is the most accurate method of metallic pipeline condition assessment available.

The MFL inspection had some challenges and was an excellent research and development experience for MFL services. One challenge came in the geometric analysis, in that the tool was not specifically designed for measuring the core of the pipe. The tool was originally designed for oil and gas pipelines to find potential obstructions in the pipe before a condition assessment, as opposed to structural anomalies. Although the geometry tool worked well in straight areas of the pipeline, it had trouble adapting to turns and bends because of its mechanical build, which slowed down the overall process. In the future, laser profiling will be used to get more efficient and accurate measurements in the geometric analysis. Another challenge came from the landscape in San Diego, which is mountainous, and required the MFL tool to be pulled up and lowered down hills using a winch. This process is extremely difficult, slow, and time consuming.

During analysis, there were a few specific challenges. Because the MFL tool collects a huge amount of data, it is difficult to transfer this data to analysts over the Internet. To mitigate this, the data had to be uploaded to an external hard drive and shipped express. This created a longer lag time between the inspection and the analysis of the data collected. There is also a challenge in locating the actual spot of the corrosion based on the data. Because the concrete that coats the inside of a metallic pipe must be chipped away to see external corrosion, any small discrepancy can make the actual corrosion hard to find. Because of this, the inspection staff needed to be extremely careful during analysis before identifying areas to verify.

The results of the MFL inspection were positive, showing that the SDCWA did not need to replace the entire pipeline at a high capital cost, but instead make specific repairs to handle the additional flow from the desalination plant. The inspection was also beneficial to the water industry showing that MFL can be used on cement lined steel pipe. It was discovered that the tool was able to successfully inspect through 1-in. mortar thickness. Out of the approximately 892 pipes inspected, seven pipes exhibited areas of broad metal loss due to corrosion. Several of these pipes were identified by the client as suitable candidates for verification of metal loss from the results obtained with the MFL inspection.

12.5 PINELLAS COUNTY UTILITIES, FLORIDA: SUBAQUEOUS PIPELINE INSPECTION

Project Owner Pinellas County Utilities (Mike Sweet, P.E., Bill Gill, P.E.)	**Technology** Ultrasonic, Magnetic Flux Leakage, Remote Field Eddy Current, CCTV, and Soil Corrosion Surveys
Pipeline Material(s) Ductile Iron Pipe, Cast Iron Pipe	**Project Date** 02/2008 to 04/2009
Consultant Atkins Kenneth E. Wilson, P.E. Ricardo Borromeo, P.E.	**Project Cost** $589,000
Contractor Mears Group J.R. Giese Operations, LLC	**Owner Contact** Mike Sweet, P.E. Utilities Director Pinellas County Utilities 14 South Fort Harrison Ave. Clearwater, FL 33756

12.5.1 Project Overview and Justification

Pinellas County is located on the Central Gulf coast of Florida. Fig. 12-12 illustrates the general location of Pinellas County within the State of Florida. Pinellas County Utilities (PCU) provides both potable water and wastewater services to all parts of the county. The coastal nature of Pinellas County requires many of the collection, distribution, and transmission mains to cross water bodies via subaqueous crossings.

These subaqueous crossings traverse the Intracoastal Waterway, bayous, canals, and creeks throughout Pinellas County. The subaqueous crossings consist of both landward and submerged segments that present constraints associated with accessibility for external inspections. The nature of the subaqueous crossings is that the majority of the crossings are underwater, and in many cases the body of water being crossed is environmentally sensitive to construction and maintenance activities. Most of the crossings have been in service for more than 30 years and are constructed with either ductile or cast iron pipe. Fourteen major water and 10 major wastewater pipeline crossings exist.

PCU was seeking to develop a program that would provide continuous inspection data along the entire length of each subaqueous crossing (both landward and submerged segments). The intent of the project was to develop an organized and systematic subaqueous crossing inspection program, conduct the inspection program, and document the findings and results of the inspections. Based upon the results of the inspection program, a repair and replacement program would then be developed.

The physical configuration and locations of the subaqueous crossings make continuous physical inspections difficult. The program envisioned

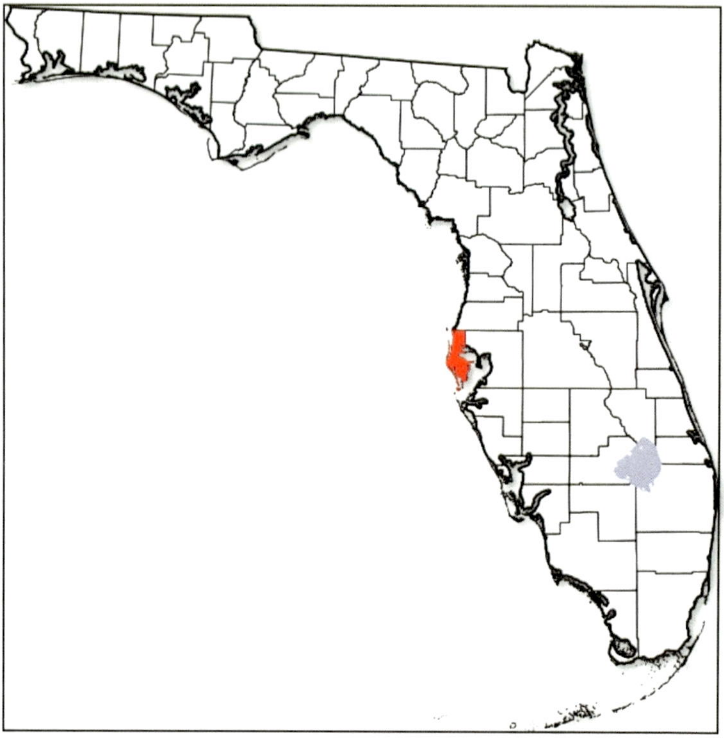

Fig. 12-12. Pinellas County, FL
Source: Courtesy of Atkins

using innovative smart pig inspection techniques to inspect the subaqueous crossings.

The first phase of the program was to develop a planning document for the subsequent inspection program that would

- Identify and describe the subaqueous crossings to be inspected,
- Evaluate available nondestructive inspection technologies,
- Discuss operational and logistical constraints associated with the inspection operations,
- Develop an initial prioritization ranking for the subject subaqueous crossings,
- Identify capital improvements required for the physical inspection program,
- Provide information about the program to various nondestructive inspection vendors, and
- Estimate probable costs for the physical inspection program.

The program described would be innovative for municipal water infrastructure facilities. The technologies described would be relatively new to larger transmission mains and in some cases untested with cast and ductile iron pipelines. Because the application of these technologies was new and innovative, care and caution was to be used to develop an effective nondestructive inspection program that provided continuous pipe condition data for PCU's potable water subaqueous crossings.

As mentioned previously, 24 cast and ductile iron potable water and wastewater subaqueous crossings were planned to be inspected by a nondestructive inspection method using smart pig technology.

The subaqueous crossings range from 125 ft to 4,700 ft in length and from 12 to 30 in. in diameter. Several of the crossings involve the Intracoastal Waterway. The marine environment coupled with maritime navigational issues makes the condition and configuration of these crossings critical. Damage to the marine environment is possible if the crossing is compromised by long-term corrosion or incidents with vessels.

12.5.2 Technologies

Three smart pig inspection technologies were evaluated for use on the subaqueous crossings:

- Ultrasonic (UT) inspection technology,
- Magnetic flux leakage (MFL) inspection technology, and
- Remote field eddy current (RFEC) inspection technology.

UT Inspection Technology UT technology uses high frequency sound waves to nondestructively inspect and measure the pipe wall thickness. Based upon the reflectivity of the sound off the pipe walls and its attenuation through the pipe walls, UT can measure the pipe wall thickness and identify anomalies in the pipe as the pig moves along the pipeline. UT has made significant technological advancements over the past few years and is considered an accurate and viable inspection method for steel pipe. Its use for cast and ductile iron pipe is much more limited.

General Electric (GE) has developed both UT and MFL smart pigs. A recent UT tool that GE has developed is the UltraScan WM tool. Unlike MFL tools, the UltraScan WM UT tool is capable of determining the location and dimensions of defects in the pipe wall. The UT signal is directed radially into the pipe to measure wall thickness. The UltraScan WM tool uses a multisensor towed array of UT transducers that provide full circumferential and axial coverage of the pipe wall. Based upon these measurements, the location, depth, and extent of both internal and external metal loss from the pipe wall can be determined. The tool requires a liquid coupling medium such as oil, refined petroleum products, or water.

Most of the current UT experience involves inspecting steel pipe for the petroleum industry. GE tested a representative sample of cast and ductile iron pipe to confirm that the UltraScan tool is applicable and to calibrate the tool to cast and ductile iron pipe. This testing, however, was not successful and therefore this technology was not considered further.

MFL Inspection Technology MFL nondestructive inspection technology uses changes in a generated magnetic field as it passes through the pipe to determine pipe wall thicknesses and to identify anomalies in the pipe. MFL was the first technology used in nondestructive inspection of steel pipe lines for the petroleum industry. MFL uses an arrangement of magnets in the smart pig that generates a steady direct current magnetic field into the pipe walls so that the magnetic field travels along the axis of the pipe. If the magnetic field is strong enough, a small amount of it will exit the pipe wall wherever a corrosion pit or similar defect is present on the pipe wall. This leakage field can then be detected by sensors on the smart pig. MFL is used for inspecting steel pipe lines and has not been used for the inspection of cast and ductile iron pipes. Discussions with GE and other MFL contractors have indicated that because cast and ductile iron pipes are more porous on a molecular level than steel pipe, it is very doubtful that MFL is applicable to cast and ductile iron pipe inspection. Initial discussions with GE revealed that samples of cast and ductile iron pipes of each diameter to be inspected need to be smart pig calibrated. The MFL smart pig is required to have close contact with the interior pipe wall to be effective. This contact is strengthened by the induced magnetic field between the smart pig and the pipe wall. As a result, the passage of the smart pig through pipe may cause the pig to scrape off linings, debris, or tuberculation within the pipe. Given the age of Pinellas County's subaqueous crossings, the use of MFL could cause damage to the crossings from the inspection activities and therefore was dropped from consideration.

RFEC Inspection Technology The RFEC technique was pioneered in the 1950s by the Shell Development Company. The first application of RFEC was in the petroleum industry similar to MFL and UT. In the 1980s, university research more fully developed the theoretical models describing the RFEC phenomenon. Based on this advanced research, new analytical methods were developed that enabled a greater range of anomalies and defects to be detected and sized. The RFEC method is now a well-established method for inspecting ferromagnetic tubes such as those in heat exchangers and boilers.

A RFEC smart pig typically has two coils, an exciter, and a detector. The smart pig is inserted into the cast or ductile iron pipe via a launching facility, and the exciter coil is energized with low frequency alternating current. An alternating magnetic field is created that travels along the interior of the pipe and is quickly attenuated over approximately 2.5 pipe diameters. The created magnetic field also passes outside of the pipe wall at the exciter coil and travels through the soil or air along the pipe exterior and passes back into the pipe wall beyond three pipe diameters from where it began (at the exciter coil). This external magnetic field is measured by the detector coil when it passes back into the pipe. The magnetic field passes through the pipe wall two times. Each passage through the pipe wall reduces the strength of the signal, and the travel time is increased. If an anomaly is present, it affects the strength of the signal, and the travel time of the field. The change in signal strength results in a change in the field's amplitude, and the change in travel time results in a phase shift of the field. With proper instrumentation and calibration, the changes in amplitude and phase can be correlated with pipe wall thickness loss. Fig. 12-13 illustrates the RFEC process.

The application of RFEC nondestructive inspection technology for the inspection of cast and ductile iron pipes has the following practical advantages:

- The ability to penetrate thick walled homogenous pipe;
- Equally sensitive to interior and exterior pipe wall anomalies;
- The ability to detect graphitization of the iron;
- The ability to test through mortar or protective pipe linings and internal scale;
- The ability to identify line valves, joints, tees, and sleeves;
- Does not require contact between the smart pig and the interior pipe wall surface; no liquid coupling medium required as with UT; and
- The inspections can be performed relatively rapid and constant pig speed is not critical.

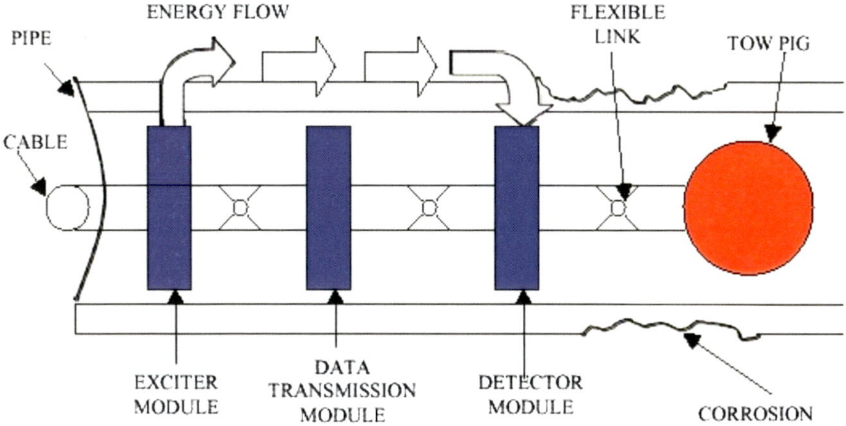

Fig. 12-13. RFEC process and component
Source: Hydroscope

Similar to the other smart pigs identified, the RFEC smart pigs can negotiate bends and fittings. The RFEC smart pigs are tethered and are not free swimming. Some are limited to 3,000 ft per inspection run due to cable limitations, and most inspections have been with 6 in. and 8 in. potable water mains. Many of the subaqueous crossings that are to be inspected are greater than 16 in. in diameter. In these cases, either another type of smart pig inspection will be required or RFEC devices will need to develop new smart pigs capable of larger diameters.

12.5.3 Evaluation of Inspection Technologies

Each of the three smart pig nondestructive inspection methods were evaluated for their suitability to the inspection of Pinellas County's cast and ductile iron potable water subaqueous crossings. Each technology has advantages and disadvantages for the inspection of the subaqueous crossings. In some cases the actual inspection methodology may not be applicable to cast and ductile iron pipes, and in other cases the pipe diameter is beyond the current limits of the respective inspection technology. Table 12-1 summarizes the advantages and disadvantages of the MFL, UT, and RFEC smart pig inspection technologies.

Based upon the technology review, both UT and MFL are not applicable for the inspection of cast and ductile iron pipe and will not be considered further. Additionally, the tight wall to sensor fit required for the MFL inspection is of concern because the interior condition of the pipes is not known at this time. The use of MFL could damage a severely corroded or damaged pipe.

As shown in Table 12-1, the RFEC technology has a good record of service for cast and ductile iron pipe inspections. The UT technology appeared to be applicable for cast and ductile iron pipe inspections, however, factory tests proved otherwise.

Based upon cost information provided by GE, the opinion of probable cost for the smart pig inspection of the 14 potable water subaqueous crossings was approximately $925,000. This was above PCU's budget for the potable water subaqueous crossings inspection program. This opinion of probable cost also did not include the cost of new pig launching and retrieval facilities required. The cost of the smart pig inspection program was prohibitive based on PCU's budget. At the time, it was decided to conduct a pilot test using the RFEC technology to determine its accuracy and cost-effectiveness for future use.

12.5.4 Pilot Program

The pilot program was conducted on an abandoned 12 in. cast iron force main that had recently failed and been replaced with a new HDPE pipe.

Table 12-1. Summary of Nondestructive Inspection Methods Advantages and Disadvantages

Nondestructive Inspection Technology	Advantages	Disadvantages
Magnetic Flux Leakage	Good record of service with steel pipes. Free swimming pigs. Can inspect large diameter pipes of infinite length.	May not work on cast or ductile iron pipe. Requires the pipe interior (walls) to be very clean. Magnetic field causes the pig to fit very tightly to the interior walls of the pipe. May scrape off linings or damage pipe during inspection process. Requires pig launching and retrieval facilities.
Ultrasonic	Good record of service with steel pipes. Free-swimming pigs. Can inspect large diameter pipes of infinite length. Recent improvements in ultrasonic technology available. Multiple firms available to do the inspection work. Pig fits loosely inside of pipe. Can determine and identify interior and exterior pipe wall anomalies and defects.	Not proven with cast or ductile iron pipe. It is believed that with proper calibration that it will work properly for cast and ductile iron pipe. Requires the pipe interior (walls) to be clean. Requires pig launching and retrieval facilities. Minimum bend radius is 1.5 × diameter.

(Continued)

Table 12-1. Summary of Nondestructive Inspection Methods Advantages and Disadvantages *(Continued)*

Nondestructive Inspection Technology	Advantages	Disadvantages
Remote Field Eddy Current	Good record of service with cast and ductile iron pipes. Proven technology for cast and ductile iron pipe inspections. Pig fits loosely inside of pipe. Pigs are launched and retrieved from the same location. Can use existing facilities (e.g. fire hydrants) to access the pipe. No pig retrieval facilities are required. Can determine and identify interior and exterior pipe wall anomalies and defects.	Tethered pigs with limited inspection length (maximum 3,000 ft). Limited number of firms providing inspection (Hydroscope appears to be the only vendor). Maximum pipe diameter that can be inspected is 16 in.

The cast iron main crosses the Intracoastal Waterway in Indian Rocks Beach and is approximately 945 ft in length. The smart pig selected for use was a customized 12 in. inspection tool based on Russell NDE Systems (RNDE) See Snake line of tools.

The pilot program was divided into four phases as follows:

- Phase 1—Planning and Logistics,
- Phase 2—Pipe Preparation and Investigation,
- Phase 3—Pipe Inspection with Smart Pig, and
- Phase 4—Final Recommendations.

Phase 1 established the scope of the project and set the project schedule.

12.5.5 Phase 2—Pipe Preparation and Investigation

Phase 2 began on March 31 and was completed on April 13, 2008. Work progressed at both ends of the subaqueous pipeline and consisted of the following major items:

- Remove 20-ft section of existing 12 in. cast iron pipe and ship to lab for calibration.
- Install pigging stations.
- Flush 12 in. cast iron pipe for two hours. Swabbing pig pushed through pipe with water from west to east.
- Connect a hard ball pig. The pig was pulled through the pipe with an 11 in. sizing plate.
- Swab pig pushed through line for further cleaning, and then push through several times with air to remove water in preparation for televised inspection.
- Televise pipe from the west side, moving east until the pipe slope began to climb upward. The camera was then moved to the east side to complete televising the line. A 4-ft section of cracked pipe was discovered 90 ft from the east end of the pipeline.
- Backfill and restore all excavations.

The initial investigative work found that the outside diameter of the pipe measured 13.2 in. with a wall thickness of 0.38 in. to 0.49 in. so it was believed to be ductile iron pipe (class 52/56). However, after pipe removal and further examination, the pipe was determined to be cast iron class 21 with an average wall thickness of 0.44 in.

The cleaning and sizing pigs indicated no obstructions or defects that would prevent further testing. Review of the video files from the television inspection indicated a substantial amount of pipeline in good condition.

The results of the investigative work indicated the 12-in. cast iron pipe was in acceptable condition to allow further testing with the tool (Fig. 12-14).

Fig. 12-14. See Snake smart pig on launch tray provided by Mears Group in preparation of launch
Source: Courtesy of Atkins

12.5.6 Pipe Calibration

The purpose of the extracted 20-ft section was to provide a representative sample of the force main, which could then be used to calibrate the tool. It is important that the tool was calibrated on pipe from the same production run as the actual line, so that optimal tool settings could be determined in preparation for the Phase 3 inspection.

The sample pipe was swabbed and fitted with a PVC liner. After installation of the liner, defects were machined into a single section of the sample pipe (Fig. 12-15), and a 12-in. tool was pulled through the pipe at various settings. Calibration values were obtained. Besides the artificial calibration defects, the data also revealed extensive natural pitting present in the pipe. The measured depths varied from 40% to 94.7% of the thickness.

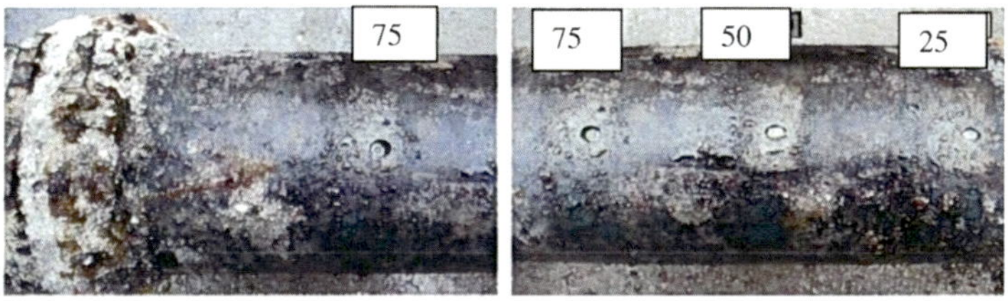

Fig. 12-15. Machined calibration reference defects
Source: Courtesy of Atkins

In conclusion, the See Snake tool was found to have no problems penetrating the cast iron of the 12-in. force main. Data quality was very good and revealed both artificial and natural defects. The presence of the hard deposits on the exterior of the pipe had no noticeable effect on the defect signals measured by the tool.

12.5.7 Phase 3—Pipe Inspection with Smart Pig

The purpose of the 12-in. Indian Rock Beach pipe inspection was twofold:

1. To assess the condition of the 12-in. line and determine the suitability for re-commissioning.
2. To prove the inspection tool worked for concrete lined cast iron force mains.

Operation

Equipment In preparation for the inspection, the contractor fed a steel tagline through the 12-in. force main, running from the east shore where the winch was set up to the west shore. With the tagline winch on the east shore, the tool winch was set up on the other side (Fig. 12-16). The inspection tool was attached to both winches allowing it to be pulled in both directions. The tool is self-contained and does not require to be powered through a wireline. It has a hard diameter of 10.25 in. and an overall length of 97 in.

Inspection The tool was placed in the west end of the pipe (exciter end first) and positioned with the detectors just outside of the pipe prior to the pull beginning, making the edge of the pipe the datum point. The tagline winch began pulling the tool toward the east on June 3 at a speed of 20 ft/min. The inspection took approximately 50 min and was stopped on the east end when the swivel head connector had just come out of the pipe.

Fig. 12-16. Tagline with tool attached
Source: Courtesy of Atkins

Field Data During the analysis of the field data from the 12-in. force main, it was observed that there appeared to be distinct zones in the data. The first 38 ft on the west shore and the last 88 ft on the east shore appeared to be thinner than the 816 ft in the middle. Initially it was believed that the thinner zones indicated areas of extensive corrosion. However, the zones appeared to start at bell and spigot connections. In addition, the cast iron pipes in the middle zone had an average length of 18.3 ft, whereas the joint segments on the east shore had a length of 17.9 ft. The video inspection indicated tight fitting joints on the east and west shores and loose fitting bell and spigot joints in the subaqueous section.

Based on the data, it appears likely that during the construction of the 12-in. force main at least two different kinds of cast iron pipe were used.

Analysis The first upstream 37.5 ft of cast iron force main exhibited significant pitting. The pitting was mostly located at the six o'clock position, with some pitting observed at other positions around the circumference.

A local wall loss was indicated at 15.5 ft. Although it is approximately 60% deep, it is very likely that it has a small center core that is deeper and possibly almost through. Also notable was a cluster of defects from 3.5 ft to 12 ft. The defects appeared to be approximately 40% deep.

After approximately 37.5 ft, the tool entered the subaqueous section of the line. This section appears to be in good shape.

A total of four indications were deemed noteworthy. At all four locations, the signals behaved very similarly and the distance to the nearest bell and spigot joint was approximately the same. It is therefore possible that these signal indications were related to the installation of the line, perhaps corresponding to steel-reinforced anchor points or other metallic features. Archival information on the installation procedure for the subaqueous section of the line would have been helpful in understanding the nature of these indications. Without historical record and/or a proper calibration pipe, it was impossible to establish the nature of these indications more accurately.

The easternmost 88 ft of the data showed extensive pitting. This section of the line was in really poor shape, with numerous through-hole indications. The largest wall loss indication extended over 4.5 ft in the axial direction. The pitting on the east shore was very heavy, with many indications sizing as through holes or close.

Conclusions The data gathering was successful, and upon analysis of the results, there appears to be three distinct sections along the length of the main with different levels of remaining pipe integrity:

1. The westernmost 37.5 ft. The pipe joints in this section match the calibration pipe and show numerous pitting indications mostly at the six o'clock position. The majority of the indications are around 50% in depth, with a couple of isolated flaws sizing deeper.
2. The section underneath the waterway (816 ft). The data from this section are significantly different from the data from the previous section. It exhibited no evidence of major corrosion and points toward the presence of additional metal wall thickness (compared with the calibration pipe). A few anomalous signals were noted, which were possibly related to the installation of the line. Archival information on the installation procedure would be required to establish the true nature of these anomalous indications.
3. The easternmost 88 ft. Pipe joints in this section of the line matched the calibration pipe. The section was in really poor shape, with numerous through-hole indications. The largest wall loss indication extended over 4.5 ft in the axial direction.

Overview of Projects To verify the accuracy of the technology, as well as evaluate the corrosion that had occurred on the pipe structure and the conclusions reached, a series of further tests were performed. These tests were direct examination test, inframetrix broadband EM survey, soil corrosion survey, and a laboratory materials examination.

On March 10 and 11, 2009, the pipeline was excavated and tests were performed and samples taken. All tests supported the corrosion locations and severity of corrosion as concluded by the data.

12.5.8 Phase 4—Recommendations

As anticipated, the results of the inspection showed that the section of pipe underneath the waterway was in relatively good condition, but the entry and exit sections above the waterline were not. The technology was found to work on cement lined cast iron pipe. Other technologies exist, but they are either visual, acoustic, or laser mapping that do not assess the structural integrity on these pipe materials. The See Snake technology gives the entire wall thickness remaining and the obvious problem locations, thus providing the opportunity to calculate the remaining wall left and remaining life of the structure. However, at the end of the project, inspection tools were not available larger than 16-in. but were being developed. Based upon discussions with RNDE on the development of larger tools, a four-year program was recommended to PCU for inspection of all 24 of its subaqueous crossings.

12.6 DALLAS WATER UTILITIES, TEXAS: WATER MANAGEMENT PROGRAM

Project Owner
Dallas Water Utilities

Pipeline Material(s)
Prestressed Concrete Cylinder Pipe, Ductile Iron Pipe, Cast Iron Pipe

Consultant
Pure Technologies

Technology
Sahara Leak Detection System

Project Date
08/2004 to present

Project Cost:
~$2.5M

Owner Contact
Randy Payton
Interim Assistant Director
1500 Marilla Dr Ste 4AN, Dallas,TX 75201
(214) 651-1441

Source: Pure Technologies U.S.

Pure Technologies has been working with the City of Dallas and the Dallas Water Utilities (DWU) in a long-term water management program, including EM inspection since 2000 and leak detection since 2004. DWU has been using Sahara for leak detection on its large-diameter water transmission mains that range from 12 in. to 84 in. The transmission mains are made up mostly of PCC, but also feature cast iron pipe and ductile iron pipe. To date, approximately 86 mi have been inspected.

Historically, large-diameter water transmission mains in Dallas have had a higher potential of developing leaks in the summer because of the extreme heat and dryness. The soil gets extremely dry and shifts the pipes slightly, causing leaks and water loss. In 2004, the City of Dallas performed an internal study of its large-diameter leak detection and found that it needed new technologies to improve efficiency. The previous system used small-diameter leak detection methods that are generally ineffective on larger pipes because of the depth of the pipe and lack of close access points. The study found that when using small-diameter methods the leak signal weakens in large-diameter pipes and that location accuracy decreased by 50% in nonmetallic pipes.

Sahara is a nondestructive leak detection tool that is pulled by the water flow by a small drag chute (Fig. 12-17). When the sensor is inserted into a tap, it remains tethered to the surface to allow for confirmation of suspected leaks. The sensor is also tracked along the surface, allowing for precise marking of leaks when they are located. In the annual preplanning stage, the DWU identifies the access points for the inspection based on its knowledge of the process from previous inspections—there are usually approximately 30 2-in. access points each year. The inspection is usually done during the summer months when most of the leaks are developing,

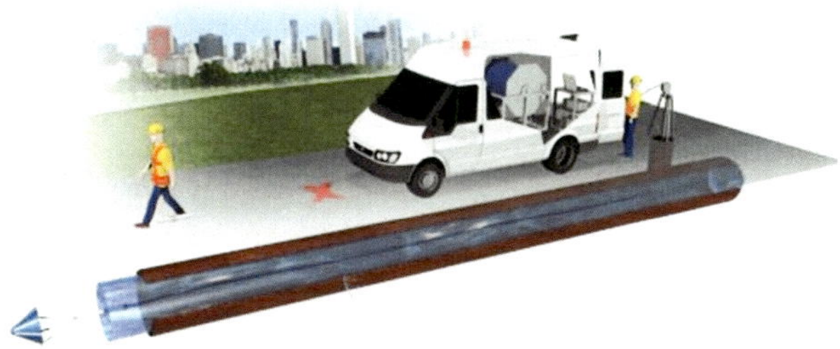

Fig. 12-17. Sahara leak detection system
Source: Pure Technologies U.S.

and higher volumes in the pipelines allow longer distances to be inspected. DWU currently inspects approximately 30 mi of pipe per year.

During the inspections, there is significant traffic control required because the tool often goes through busy streets or downtown Dallas. To avoid major commuter disruptions, the City of Dallas will reroute traffic and thoroughly plan the inspection to avoid high traffic times—for example, starting the inspection in the midmorning when traffic slows as opposed to during morning rush hour. An advantage to this tool is that it is contained in one truck that can be quickly set up to perform work at street manholes, which was crucial for the City (Fig. 12-18).

During the inspection, there are unavoidable environmental challenges that require adjustments. Sometimes the water main will run under a busy highway or an environmental obstacle like a river, making it impossible for the staff member on the ground to track the tool and mark exact leak locations. In this case, the operator needs to review potential leaks more closely by moving the tool back and forth to determine the exact location. The tool also requires close control of the flow rate to ensure that it can reach its full intended distance. After years of leak inspection/detection, DWU has become experts at controlling the flow rate and identifying precise access points that maximize the length of the tool's inspection capabilities.

The inspection allows for real time location of leaks, cracks, and gas pockets—meaning there is very little review of data after the inspection. Because the tool can be moved back and forth when a leak is suspected, it is

Fig. 12-18. Sahara leak detection system manhole setup
Source: Pure Technologies U.S.

usually possible to locate leaks immediately. If there are environmental restrictions preventing the ground operator from tracking the tool and marking leaks, the machine operator will verify leak locations by moving the tool back and forth to find the location.

These inspections in Dallas have been extremely successful, locating 116 leaks in the 86 inspected miles. The estimated water savings from these leaks is approximately 7.2 million gallons per day. DWU has also seen a 17% reduction in catastrophic water main failures, likely as a result of the proactive approach to fixing leaks. The reduction in failures has reduced property loss claims and service interruptions, as well as reduced treatment and delivery costs.

During each inspection, Sahara operators mark leaks for repair. The DWU is very proactive in repairing these leaks shortly after the inspection and has repaired approximately half of the identified leaks since the conclusion of the 2012 inspection in July.

12.7 CITY OF HOUSTON DEPARTMENT OF PUBLIC WORKS AND ENGINEERING, TEXAS: REHABILITATION OF 60-YEAR-OLD CAST IRON WATER LINE

Project Owner City of Houston Department of Public Works and Engineering **Consultant** Lockwood, Andrews & Newnam **Contractor** Boyer/Miller Pipeline **Technology** Internal Joint Seals and Augering	**Project Date** 07/2006 to 04/2008 **Project Cost** $4,700,000.00 **Owner Contact** Mr. Arthur Morris, P.E. Managing Engineer 611 Walker, 15th Floor Houston, Texas 77002 hamlet.hovsepian@houstontx.gov

The City of Houston, the fourth largest city in America, is facing the same dilemma as other metropolitan areas across the country: how to do more with less funding. A recent study performed on behalf of the City of Houston Department of Public Works and Engineering (DPWE) by Lockwood, Andrews & Newnam identified a priority of projects requiring replacement and rehabilitation. In the past several years, DPWE has been experiencing problems with existing 24-, 30-, and 36-in. cast iron water lines concentrated in the vicinity of its Museum District and the Galleria Mall retail and business area. These two areas experience very high traffic volumes and are considered some of the City of Houston's most recognized places.

The 24-, 30-, and 36-in. water lines were constructed in the 1940s and 1950s and consist of cast iron material with lead caulked joints. The leak history of these lines indicates that a disproportionate amount of effort is required by DPWE forces to maintain these lines, making them a maintenance burden for the City of Houston. However, coupons cut from the existing line indicate the existing pipeline is structurally sound and free of corrosion—the problem was at the joints.

Complicating the project design efforts, the project required extensive coordination with the City of Houston, Texas Department of Transportation, and a major shopping center, requiring carefully planned traffic control measures and special provisions to minimize impacts.

12.7.1 History

Over the past several years, a growing burden to the DPWE's maintenance division has been numerous leak repair calls on existing large-diameter cast iron water lines. At the time of construction, cast iron was chosen for the 24-, 30-, and 36-in. water lines because of the pipe material's availability and strength properties (Fig. 12-19). One particular attribute of cast iron is its corrosion resistance, which makes it well-suited for Houston soils and exhibits a long service life.

The bell and spigot joints on these pipes were completed by tamping oakum into the base of the annular bell cavity and filling in the remaining

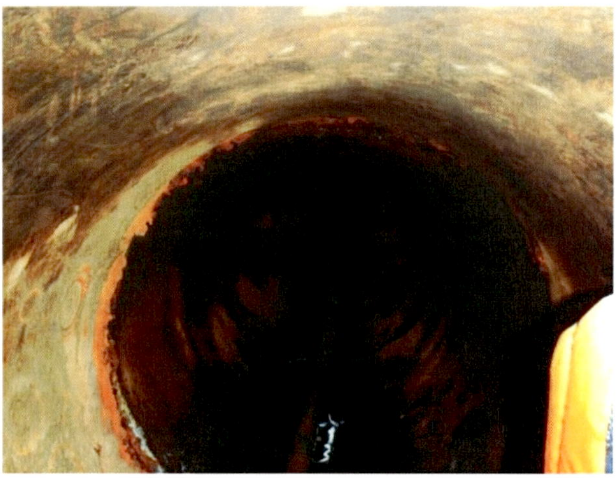

Fig. 12-19. Interior of existing cast iron water line
Source: Courtesy of Lockwood, Andrews & Newnam

space of the bell with molten lead. The joint was then caulked by tamping the solidified lead with special tools to achieve a watertight seal.

In recent years, the DPWE maintenance crews have been continually performing point repairs by recaulking the joints and installing repair clamps over the bell portion. Some joints could not receive a readily available repair clamp because the bell portions were not manufactured to today's industry standards. Any exterior repairs require a daytime excavation at the joint while impacting the water supply to surrounding areas. Local traffic patterns are affected as well due to lengthy lane closures. Recaulking joints is not only time consuming, but is only effective for sealing currently visible leaks. Due to the amount of time and cost spent annually repairing the leak, the City of Houston preferred a more proactive approach to eliminate the consistent impacts to the Galleria retail area and surrounding residential areas.

Lockwood, Andrews and Newnam (LAN) was retained by the DPWE to conduct a study to compile new information from available documentation about the existing 24-, 30-, and 36-in. water lines and to recommend necessary courses of action to improve the performance and enhance functionality of these lines. The evaluation began by collecting data from three important areas: (1) record documents, (2) repairs, and (3) condition assessments. Analysis of these data determined which sections of the 24-, 30-, and 36-in. water lines require rehabilitation.

12.7.2 Data Collection/Records Research

Records Documents Evaluating record documents assisted in determining the approximate location of the existing 24-, 30-, and 36-in. water lines and small-diameter interconnections. These documents served as a guide because it is common for realignment to occur due to unforeseen obstructions during construction. Having a complete field survey will verify the actual location of the water line. The study analyzed available as-built record drawings, the DPWE's geographic information and management system, water meter account logs, and up-to-date repair history of the water lines.

As-built record drawings for the large-diameter water lines along Westheimer and Mid Lane, Calumet, and Westpark were retrieved from the City of Houston to determine the location of pipe features, valves, and interconnections. Many of the record drawings for the interconnecting water lines were unavailable, and many changes had been made since the original construction. Therefore, confirming which customers and small-diameter lines were fed by these mains needed to be field confirmed. These interconnections were analyzed as being potentially impacted by a shutdown of the 24-, 30-, and 36-in. water lines. This analysis resulted in a need for constructing parallel distribution lines along the large-diameter cast iron water lines to supply water to these existing services.

In addition to these maps, water service meter records were reviewed for an accurate representation of potential affected customers by a water main shutdown. These meter records account only for customers receiving service directly from the main lines. In-depth research determined that numerous customers receiving water from the interconnected lines would also be affected.

The repair history of the 24-, 30-, and 36-in. water lines was also evaluated in an attempt to prioritize segments that required improvements. The information, compiled in water infrastructure replacement prioritization (WIRP) maps, represents the number of work orders issued for repairs to the 24-, 30-, and 36-in. water lines. These maps are compiled by the DPWE and provide information about the work order, including the actual location of the water line, cause for the work order, and resulting action. However, this information was only beneficial as a tool to gauge the condition of a particular stretch of water line and to prioritize areas for further study. Table 12-2 is a summary of the available WIRP data along with the installed date of the 24-, 30-, and 36-in. water lines.

This evaluation demonstrated that the 24-, 30-, and 36-in. water lines were a growing maintenance burden for the City of Houston and required at least 69 separate repairs in the last five years. The repair records also displayed the DPWE maintenance crew's description of the problem and the effort to repair the leak. More than half of the leak repairs required the line to be temporarily isolated, and almost all repairs required lanes of traffic to be blocked.

Field Survey Results

Westheimer and Mid Lane Westheimer is a varying width right-of-way section with six lanes that narrow down to four lanes to the east of IH-610. Development along Westheimer is primarily commercial that is intermixed with residential properties along the route. Mid Lane, conversely, is a 60-ft right-of-way, two-lane street that is primarily residential and is characterized by large homes and townhomes. Landscaping and large trees, which

Table 12-2. Summary of Repair History for Large Diameter Mains

Street	Size	Installed Date	Repair Work Orders (2003–2008)
Calumet	30 in.	1944	13
Westpark	36 in.	1946	15
Westheimer and Mid Lane	24 in.	1953	41

form a canopy over portions of the roadway, characterize the Mid Lane right-of-way. Traffic along Westheimer is considerable due to the highly publicized Galleria Mall located within close proximity of the project site. Impacting current traffic patterns in this area with extensive construction activities would significantly reduce the expected revenues from this highly publicized area. Mitigating was extensively evaluated. Mid Lane's traffic consists of normal commuter travel during the daytime hours. Construction activities for Mid Lane had little impact on the daily traffic pattern due to numerous side streets to which the traffic was diverted. Overhead and underground utility services extend along both sides of the right-of-way for Westheimer and Mid Lane, which provided a host of conflicts during design and construction.

Calumet Calumet is an 80-ft right-of-way, two-lane boulevard section that is primarily a residential neighborhood near the Museum District of Houston. This area of Houston, just like the Galleria area, is frequented by numerous tourists on a daily basis. Impacting this area would cause a decrease in tourist visits to the well-renowned area. Mitigating this impact was extensively evaluated. Overhead and underground utility services crossed, as well as extended along, both sides of Calumet's right-of-way, which also provided a host of conflicts during design and construction. In addition to the utility services, the possibility of a shutdown of the 30-in. water line affecting a hospital and an academia school, located within five blocks, was a major concern.

Westpark Westpark is an 80-ft right-of-way, four-lane highway with a continuous left turn lane in the center. A Metro Transit Authority easement and Reliant electrical easement extend along the eastbound lane for the entire length of the alignment. Commercial properties extend along the north right-of-way of Westpark, whereas along the south right-of-way high voltage power lines and residential properties extend along the alignment. In addition to these properties, a hospital and an academia school are within close proximity to Westpark and therefore could have been affected by the shutdown of the 36-in. water line.

12.7.3 Condition Assessment

In addition to the analysis of record documents and field survey, evaluating the existing condition of the water lines assisted in determining the correct course of action. To accomplish this evaluation, sections of the water line needed to be exposed for an external evaluation. Even though the whole line could not be exposed, this intermittent evaluation led the authors to accept that the remaining line's structural integrity would be consistent with the condition found. Evaluation of the 24-, 30-, and 36-in.

water lines' existing condition revealed a structurally sound pipe with minimal visible signs of deterioration.

To evaluate the interior condition of the water lines, engineers inspected the pipe exposed during several repairs, and coupons were cut from several locations using a tapping sleeve (Fig. 12-20). Coupons taken from the 24-, 30-, and 36-in. water lines revealed that these lines consisted of approximately 1-in. thick cast iron with roughly 1/8-in. lining of mortar. This mortar lining was assumed to extend the length of the project limits. Evaluations of the coupons indicated that the mortar lining is in good condition, and the pipe did not show signs of corrosion, either inside or outside. Therefore, it was concluded that the 24-, 30-, and 36-in. water lines had no issue with performance capabilities.

12.7.4 Evaluated Alternatives

The only options available to remediate the leaks along Calumet, Westpark, Westheimer, and Mid Lane were complete replacement or rehabilitation. To determine which option was best, the following criteria were evaluated:

- Feasible construction,
- Low associated construction costs,
- Expedited contract time, and
- Minimal impact to current traffic patterns.

Fig. 12-20. Coupon removed from the 24 in. cast iron pipe
Source: Courtesy of Lockwood, Andrews & Newnam

Complete replacement of the water line was removed from consideration based on the following:

- High associated construction costs,
- Impacts on commercial and residential traffic, and
- Lengthy contract time.

Rehabilitation was then divided into two different methods, internal or external, for further analysis. External rehabilitation requires excavating each joint for caulking the lead into the bell cavity. This method would be a continuation of the DPWE's effort in repairing leaks. However, the construction time involved, as well as the costs associated with this effort, was comparable to the efforts required to replace the entire water line. This method of rehabilitation would have also required an extensive use of traffic control devices that would severely impact normal commercial and residential traffic. This method is better used for point repairs on a limited number of leaks.

Internal rehabilitation uses various techniques for eliminating leaks, which include, but are not limited to, slip lining or internal seals at each of the joints. Sliplining was removed from consideration because of the following:

- Associated construction cost,
- Extensive traffic control plan,
- Lengthy main shutdown required, and
- Considerable reduction in pipe capacity.

Another method of internal rehabilitation is to install rubber seals at the joints, held in place by stainless steel bands (Fig. 12-21). This method requires that an individual enter through an access manway into the pipe. To install the seal, the individual would be required to remove some of the

Fig. 12-21. Installation of an internal joint seal
Source: Courtesy of Lockwood, Andrews & Newnam

mortar lining of the pipe to obtain a watertight seal. Held in place by a stainless steel retaining band, the seal creates a low profile that prevents an increase in head losses within the pipe. The seals can also be tested to ensure that a watertight seal is obtained.

12.7.5 Recommended Alternative

LAN recommended rehabilitation of the 24-, 30-, and 36-in. water lines based on the evaluation of the field survey data and existing condition of the pipe. Rehabilitating the joints, in particular, would reduce the probability of leaks from recurring. Internal joint seals, in this case, WEKO-Seals supplied and installed by Miller Pipeline, were chosen based on the method's capabilities of achieving the evaluation criteria. Internal joint seals were chosen based on their low installed profile, which prevents large head losses within the pipe; the testing requirements to be achieved before the line is returned to service; and lower installed costs. Cost analysis showed that rehabilitation using joint seals, including construction of new distribution lines needed, would cost approximately one-third the price of installation of a new large-diameter transmission line.

Table 12-3 displays the recommended rehabilitation limits of the 24-, 30-, and 36-in. water lines.

12.7.6 Challenges

The important challenges that were recognized and incorporated into the design phase of this project were

- Traffic control,
- Maintaining local water service,
- Access into the pipe, and
- Leaking in-line valves.

Specific traffic control design was prepared to mitigate impacts to traffic flow while reducing driver confusion and providing a safe work environment for the construction workers. One example of an effective traffic control plan was proposing the work to occur during the nighttime hours for the high traffic volume areas and avoiding work during peak shopping seasons around the Galleria Mall.

Table 12-3. Rehabilitation Lengths for 24-, 30-, and 36-in. Water Mains

Street	Size	Rehab. Length (linear ft)
Westheimer and Mid Lane	24 in.	8,000
Westpark	36 in.	5,700
Calumet	30 in.	3,300

To rehabilitate the lines, they needed to be isolated and removed from service. At the time, they operated as both transmission and distribution lines. The primary purpose for the lines was to carry large amounts of water from the DPWE's Southwest Pump Station; however, over the years, many residents, businesses, and connections to local distribution lines had been added. To maintain the current water supply to customers during the rehabilitation and permanently thereafter, approximately 12,500 linear ft of small-diameter (8- and 12-in.) distribution lines were constructed within the current right-of-way of the existing large-diameter lines, as a parallel system. The small-diameter water lines were installed by augering, to further minimize disruption associated with open-cut. This method involves minimal surface impact on the existing right-of-way, as well as limiting traffic disruption.

Existing services along the alignment were transferred to the new distribution lines prior to isolating the large-diameter lines for rehabilitation activities. With the services transferred to the new distribution lines and the rehabilitation of existing lines complete, the 24-, 30-, and 36-in. water lines were converted into dedicated transmission mains to be used with potential future expansions.

To install joint seals, work crews needed to gain entry into the pipe, which was not constructed with access manways or other outlets. Also, for safety reasons, two isolation valves or a valve and air gap needed to be provided on each side of any pipe entered. Therefore, the design included multiple new access manways (Fig. 12-22) and isolation valves to be installed prior to rehabilitation.

Fig. 12-22. New access manway in manhole on 24 in. pipe
Source: Courtesy of Lockwood, Andrews & Newnam

Another challenge presented during this project was the possibility of nonfunctioning isolation valves. The in-line valves had been in service since the inception of the 24-, 30-, and 36-in. water lines and had only been replaced or serviced as needed. With so many older valves still in place, all valves needed to be assessed for operability and capability before pipe rehabilitation could commence.

The prospect of leaking valves also limited the ability to hydrostatically test the line in accordance with standard city procedures. Therefore, two requirements were placed on the contractor. First, each joint seal would be individually tested after installation, and second, a full scale, but low pressure, hydrostatic test would be performed on the entire line. This latter test was proven successful at normal operating pressure of the line.

Construction of approximately 12,000 linear ft of 8- and 12-in. water line, transfer of 175 customer services, and installation of approximately 1,500 internal joint seals were successfully completed with little disruption to the public over a one-year duration at a cost of $4.5 million.

As of June 2016, approximately 8 years since completion of the work, no new leaks have been reported on the rehabilitated lines.

INDEX

Pages with tables, figures, and equations are indicated by *t*, *f*, and *e*